Active Landscape Photography

How can photography be transformed into an active process of investigation for landscape architecture and environmental design? The second book in Godfrey's series, *Active Landscape Photography*, presents engaged photographic methods that turn photography into a rigorous, thoughtful endeavor for the research, planning and design of landscape places.

Photography is the most ubiquitous and important form of representation in these disciplines. Yet photography is not specifically taught as a core skill within these fields. This book creates a starting point for filling this gap. Concepts and working methods from contemporary photography and critical cultural theories are contextualized into situations encountered in the daily practice of landscape architecture and environmental design. These methods can be integrated into practices in academic and professional settings or picked up and self-taught by an individual reader. Part I: Methods presents easily accessible approaches to photography, creating a core set of active skills. Part II: Practices discusses working methods of specific contemporary photographers and extrapolates their practices into common planning and design situations. Contemporary photographers presented include Richard Misrach, Dawoud Bey, Duane Michals, Latoya Ruby Frazier, Mark Klett, Sophie Calle, Joe Deal, Robert Adams, Naima Green, Bernd and Hilla Becher, Stephen Shore, David Hockney, Amy Sherald, William Christenberry, Jeff Wall, and Sohei Nishino.

Beautifully illustrated in full color with over 150 images by Godfrey, her students and contemporary photographers, this book provides both clear guidelines for a set of diverse methods and a deeper discussion about the implications of making and using photography in environmental design for professionals, academics, students and researchers.

Anne C Godfrey is an Assistant Professor in the Department of Landscape Architecture at SUNY College of Environmental Science and Forestry. Her research examines the complex relationship between photography and landscape. As a photographer and writer, she engages in contemporary multidisciplinary inquiries into landscape, perception and design. She is an award-winning educator and researcher.

Active Landscape Photography

Methods for Investigation

Anne C Godfrey

Routledge
Taylor & Francis Group
LONDON AND NEW YORK

Cover image: Anne C Godfrey. Penobscot Bay, Maine. September 2021

First published 2023
by Routledge
4 Park Square, Milton Park, Abingdon, Oxon OX14 4RN

and by Routledge
605 Third Avenue, New York, NY 10158

Routledge is an imprint of the Taylor & Francis Group, an informa business

British Library Cataloguing-in-Publication Data
A catalogue record for this book is available from the British Library

Library of Congress Cataloging-in-Publication Data
Names: Godfrey, Anne C., author.
Title: Active landscape photography : methods for investigation / Anne C. Godfrey.
Description: Abingdon, Oxon ; New York, NY : Routledge, 2023. | Series: Active landscape photography | Includes bibliographical references and index.
Identifiers: LCCN 2022011469 (print) | LCCN 2022011470 (ebook) | ISBN 9780367248420 (hbk) | ISBN 9780367248444 (pbk) | ISBN 9780429284649 (ebk)
Subjects: LCSH: Landscape architecture—Photographs. | Landscape architecture—Pictorial works. | Landscape photography.
Classification: LCC SB472 .G634 2023 (print) | LCC SB472 (ebook) | DDC 712.022/2—dc23/eng/20220722
LC record available at https://lccn.loc.gov/2022011469
LC ebook record available at https://lccn.loc.gov/2022011470

ISBN: 978-0-367-24842-0 (hbk)
ISBN: 978-0-367-24844-4 (pbk)
ISBN: 978-0-429-28464-9 (ebk)

DOI: 10.4324/9780429284649

Typeset in Futura
by Apex CoVantage, LLC

To Don

Contents

Introduction 1

Terms, Assumptions and Concepts 20

PART I
Methods 27

Method 1 Slowing Down 29

Method 2 Limits 35

Method 3 Asking Questions 43

Method 4 Not Photographing 48

Method 5 Quick Simple Things to Pay Attention to While Photographing 52

Method 6 Frame: Reframe 57

Method 7 Compositional Sets: Horizon Line, Focal Point, Near and Far 60

Method 8 Changing Point of View: Imagining and Embodying 67

Method 9 Photographing What Someone Else Sees 72

Method 10 Casual Composition: Note-Taking 76

Method 11 Time With Photographs 80

Method 12 Significance of the Unphotographed 84

PART II
Practices 87

Practice 1 Series: Narrative 89

Practice 2 Predetermined Path: Regularized Photographing 97

Practice 3 Distance: Stepping Back 104

Practice 4 Re-Photography: More Than Comparison 112

Practice 5 Community 120

Practice 6 Things Take Time: Time Takes Things 126

Practice 7 Joining: Combining 130

Practice 8 Rules: Games, Drifts, Prompts 137

Practice 9 Many Others: Many Interpretations 143

Practice 10 Narrative: Writing What You Don't See 149

Acknowledgments 157
Bibliography 159
Index 168

Introduction

WHY PHOTOGRAPHY MATTERS, WHY METHODS ACTIVATE PHOTOGRAPHY

Photography is the most important and accessible tool for design decision making in landscape architecture and the related environmental design disciplines. Photography is made and used to document, investigate and understand landscape sites and contexts. Photographs are viewed throughout the design process to assist in decision making. Photography is a daily helpmate to our work.

Photographs are combined and manipulated with an ever-increasing host of digital software tools. Photographs are used as underlayments for digital design proposals.

Photographs are collaged, morphed and animated to demonstrate design change and finished design results.

Photographs become the basis for 3D renderings and models. They can serve as locating devices for complex mapping processes. Imagery is created through photographic and data collection means to remotely sense and examine landscape characteristics, such as land cover, topography, human use, other animal use and water flows (see Figure 1). These photolike images provide understandable visual cues and are referenced for the "accuracy" of certain elements (such as in the construction of complex 3D maps).

We use photography to show our work to others. These photographs of finished work appear on websites, social media feeds, web and print media, monographs and reviews. We display photographs of finished work in our offices. We hire photographers to make photographs that specifically demonstrate the successes of our individual projects.[1]

Our work is evaluated through the viewing of photographs. Photographs are submitted to award calls. Most post-occupancy evaluations employ photographs as part of their core methodologies. Photographs are used to demonstrate historic and cultural significance. Sites are monitored with photography. Photographs are assessed to understand change over time.

Photography is the way we understand our history. We make photographs in the present that become part of the documentation of our work. We evaluate historic photographs to understand the designs of both well-known and newly recognized contributors to our disciplines. We collect design process documents in archives, including photographs made while designing and photographs made when the projects are finished. All these photographs combine to confirm the history of our discipline.

Additionally, all the photographs *not made* or kept determine the understanding of our history. Photography is selective, or we are when we make and use it. To photograph is to designate importance. By photographing, we choose to show things and simultaneously choose to not show things.

DOI: 10.4324/9780429284649-1

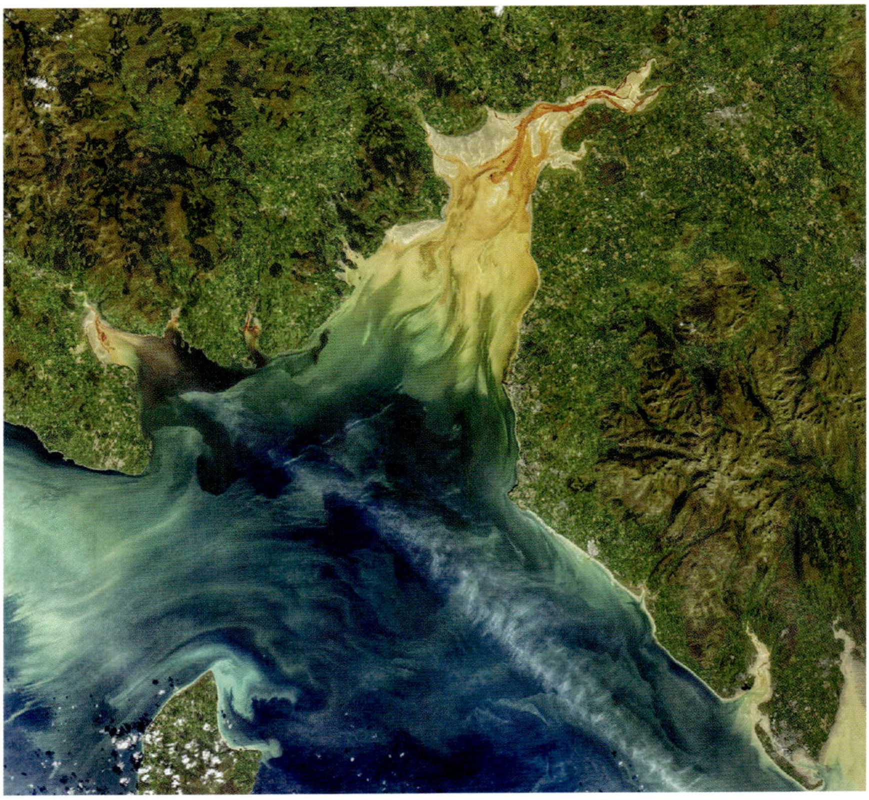

Figure 1
This image of Solway Firth in the United Kingdom was made by the Operational Land Imager (OLI) on the Landsat 8 satellite managed by the National Aeronautics and Space Administration (NASA). The OLI is a remote sensing device that measures the visible, near infrared and shortwave infrared portions of the light spectrum. The image was made to monitor sedimentation patterns on October 2, 2019. NASA's information about this image includes this narrative:

"The image above is a blend of art and science. Like a photographer who adjusts lighting in a studio, Norman Kuring of NASA's Ocean Biology group works with various software programs and color-filtering techniques to draw out the fine details in the water. The swirls and streamers in Solway Firth are real, but Kuring has separated and enhanced certain shades and tones in the data to make the sediments and dissolved organic matter stand out."

Image: Norman Kuring for NASA.

Photography is powerful because it is an image-making process that looks "a lot" like how we see with our eyes. But photography is not the same as sight. We'll unpack this further, later in this introduction. But first, let's articulate some of the common uses and conceptions of photography in landscape architecture and environmental design. Through this discussion, the usefulness of utilizing predetermined methodological approaches to photographing is introduced and explored.

PHOTOGRAPHY IS NOT CONSIDERED A SKILL TO BE LEARNED

Photography seems easy. We use it daily in the practice of landscape architecture and environmental design. Yet it is seldom closely examined or questioned because of its ubiquity and perceived ease. The first book in this series, *Active Landscape Photography:*

Theoretical Groundwork for Landscape Architecture (*ALP 1*) addresses this gap through critical discussions exploring how our use of photography influences how we understand value and design landscape places.

What's interesting about the condition of photography within our landscape-interested disciplines is that each of us is expected to photograph as part of our work, yet rarely are we expected to be formally trained in photography. The various professional degree programs train us how to draw, map, measure, design, observe and utilize diverse software – all processes that result in forms of representation. Often, specific training allows for specialization in areas of the discipline, such as public engagement, cultural landscape documentation and manipulation of geospatial data. These specialized processes also result in representations. We know photography is also a form of representation used to communicate ideas, proposals and values. Yet it is curious that photography remains a technique and skill that receives very little formalized attention in environmental design curricula despite the fact that we use photography, in all its various forms, *every day* to make design decisions.

The consumer culture around photography supports the idea that little training is necessary to take photographs. *All I have to do is point and shoot. The camera does the rest.* The advertisements for the first consumer-oriented cameras, such as Eastman Kodak's Brownie Box camera released in 1900, presented the idea that photography was easy (see Figure 2). *Everyone can do it, even a child!*[2] This type of advertising continued with the introduction of Polaroid's Land Camera in 1947[3] and later with Polaroid's One Step, in which adds read, "World's simplest. Simply press one button. Nothing to focus or set. Motor ejects the picture. It develops in minutes. . . . Simple? It's the simplest."[4]

This same rhetoric is used today. Smartphones, which now contain the most commonly used cameras, are advertised with claims such as, "Your smartphone is more than just a phone. It comes with a built-in professional-grade camera that's easy to use and there's a whole host of modes to help you capture every moment, from gorgeous selfies to stunning panoramas . . ."[5] The camera's ability to make photos so well and so easily is a primary selling point.

Thus, the idea that photography is easy perpetuates – it's something the camera does, not the person releasing the shutter or tapping the screen. Photography is not something that needs to be learned – it's something you just do.

Yet almost no other form of representation used on a daily basis in landscape architecture and environmental design is treated as something one can just pick up, do and easily achieve immediate "success." Imagine if we thought of Photoshop or AutoCAD or ArcGIS

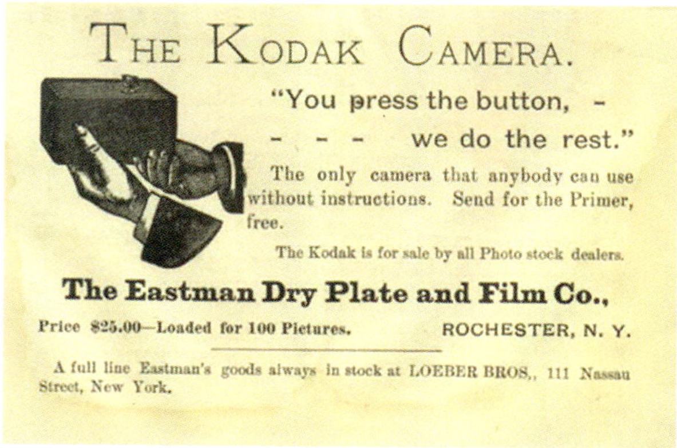

Figure 2
Early advertisements for cameras purported how easy they were to use. This ad for a Kodak camera from 1889 (a precursor to the Brownie Box camera) articulates the now commonplace assumption about photography – it's easy, and everyone can do it. "The only camera anyone can use without instructions."

Image: Loeber Bros. Kodak advertisement, public domain.

that way! Even writing is taught as a skill to be honed, through general education courses and sometimes specialized courses within environmental design degrees, while photography is simply taken for granted. This is not only a condition in environmental design, but also a generally held idea about photography, as demonstrated with the previous examples.

For environmental design, our use of photography influences how we change places. How we change places impacts people, other animals, ecological systems, economies, cultures and the physical fabric of the landscape. How we design landscapes influences everyday personal moments in a park or a bus stop or a bike lane.[6] Our designs influence water quality, storm resilience, food production, pedestrian safety, mental health and community gathering.

We use photography as a process to investigate landscape places. We make and use photography everyday as the primary assistant for gathering information to make design choices. The previous book in this series, *ALP 1*, presented a set of concepts, examples, and theories from photography and contemporary cultural studies applied to specific situations in landscape architecture and environmental design as a way to cultivate a more critical and conscious dialogue about photography. That book also highlights photography's significant roles and functions in daily decision making processes during the planning, design and management of landscape places.

This book picks up where *ALP 1* left off by presenting a set of predetermined methods for photographing that teach the reader how to actively engage a process of inquiry with photography – instead of passively pointing and shooting. These methods fill the current gap in photography training by providing structure, rigor and critical engagement practices for any photographic session that investigates the landscape.

FROM PHOTOGRAPHING EVERYTHING TO PHOTOGRAPHING WITH A PURPOSE

Engaging in predetermined methods, simple to complex, helps us actively engage in the process of photography. This is different than the default untrained point-and-shoot approach when the photographer just shows up and reacts. Photographing in a methodical way shifts photography into a process of making instead of taking.[7]

When we think of photography as a process of making, or creating, the conception of photography starts to look more like this:

> *I'm going to utilize specific methods to direct my process of photography.*

Rather than this:

> *I'm going to go shoot some photos.*

In understanding how to make this shift, let's consider some common modes for photography when photographing a design site:

> One or more of the design team, or students in the studio, start immediately photographing. The camera or phone is placed up in front of the face and quickly hundreds of photographs are taken. This mode of photographing is often about thinking that I can make a copy of what I see with my eyes with the camera, so that I can then go back to the office and look at the site (not the *photographs* of the site – notice the difference here) later.

Often this mode of photographing is joined by this thought:

> If I just take enough photographs of the site, I will get what I need.
> This leads to thinking "I can photograph everything."

Another common mode for photographing is this:

> While I am visiting a site for design, I am observing, measuring, walking, meeting with people, taking notes, diagramming and talking with colleagues. Among all of those activities, I also happen to be photographing.

Or perhaps this:

> The project manager says to a staff designer: "Could you go and remeasure (for instance) the existing conditions that are helping us locate the point of beginning for this site plan to make sure we have this right? Oh, and by the way, could you take some photographs while you are there?"

Several things are occurring in each of these instances. Photographing is done quickly. Photographing is done simultaneously with other activities. Photographing just happens and is done without much thought beyond capturing something that "might be useful later." Finally, I think that if I just photograph everything, I will successfully document the site.

The results of this kind of process[8] include (1) In general, the photography process is passive. (2) A strong connection to the process of photographing is not cultivated, because multiple things are being done at once. Awareness is not focused on photographing, so the meaning of many of the photographs gets lost. (3) So many photographs are taken that it becomes difficult to focus and sift through them in order to find ones that are helpful. This is either because the visual information becomes overwhelming or one simply does not have time to thoughtfully look through hundreds of site photographs.

These default conditions are most often predicated on the notion that the camera does all of the work. All I have to do is press the button or poke at the screen. It's a nice idea, right? Could someone, or rather something, do this work for me? The camera is my assistant. But this is not the case.

Instead, utilizing a set of predetermined methods can be the assistant. For example, a predetermined method can require that each photograph attempts to answer a specific set of questions related to the design goals of a project. For instance, a design project would be helped by understanding how existing public spaces are used in downtown Syracuse, New York. The process of photographing is directed by an explicit set of predetermined questions:

> Who is here? What are they doing? What parts of the space are they using? What parts of the space are not used? How long do they stay? Where do they enter and exit the space?

Applying a methodological approach to photography is similar to stating a thesis. For example, a research paper on downtown Syracuse cannot and is not expected to present a full discussion of all aspects of that place. Instead, a clear thesis is presented to direct the discussion toward a specific subtopic. Of course, there are multiple influencing contexts

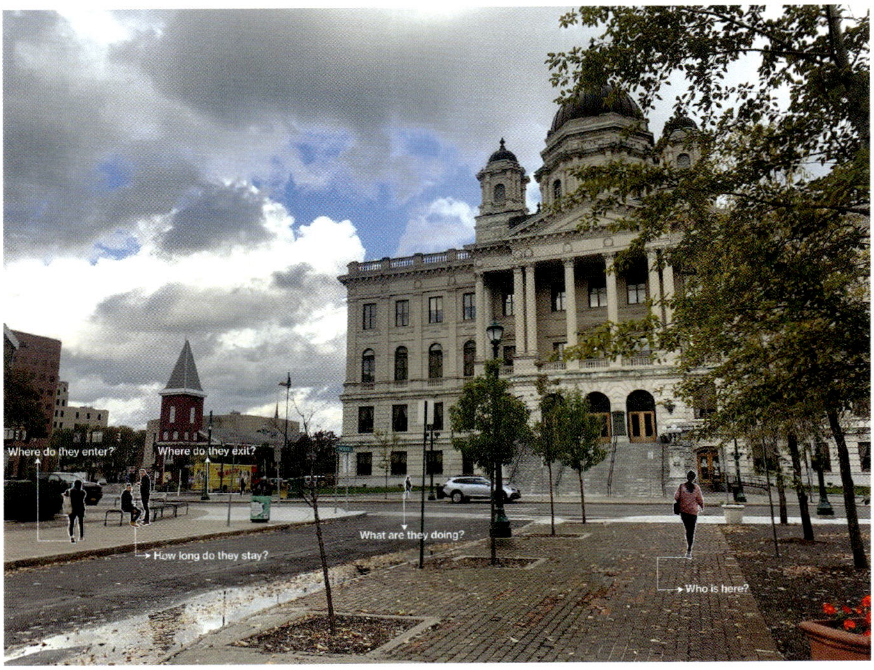

Figure 3
As a well-known public space in Syracuse, New York, Columbus Square serves many functions. Here the photograph is made to investigate current human use. The annotations help the viewer understand the motivations of the photographer/designer.

Image: Sara Constantineau, 2021.

that can be referred to, but the lens through which the place is presented is articulated by a clear specific thesis. To extend that idea to site photography, the photographic set here would focus on people in place, reflecting the focus of the thesis for the design project (see Figure 3).

I could teach a different studio in the same place. This studio could be about assessing and understanding tree health in downtown Syracuse, as a means to propose green infrastructure best practices for the city. The initial photographic process for this studio would ask these kinds of questions:

> Through visual assessment, what tree species are doing better than others? What is the typical street tree planting and infrastructure treatment? Where are there more trees or less? Are there other infrastructural elements that pair with the tree conditions? Where could more trees be planted?

These two different examples, both with specific goals, have clearly articulated agendas. When joined with, as we see here, forms of text pointing those interests out, others can better understand the focus and intended utility of the photographs (see Figure 4). This methodological approach is quite different than showing up and "photographing everything." The predetermined method assists in creating a focused set of photos made with clear intentions related to specific goals. This focused approach starts to address another common idea: that photography replicates sight. Instead, with a stated goal for photographing, the photographer starts to see how they influence the way a site is presented to others. The focus on human use or tree health results in two different sets of photographs. This process of *making* highlights how photographing does not simply copy the site.

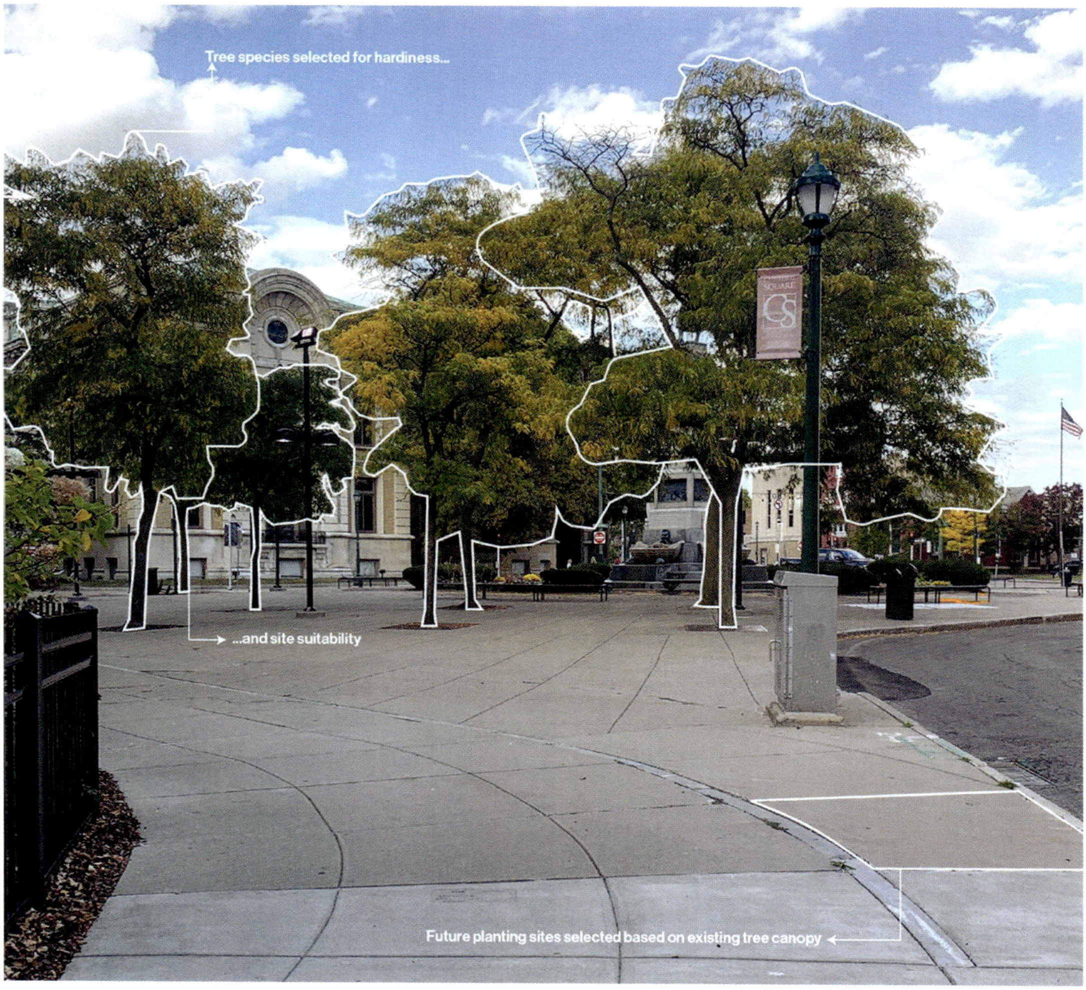

Figure 4
Here the photographer/designer is looking at the pattern of street tree planting in Columbus Square. Though this image is of the same public space as the previous image, this photograph is made with a different set of questions in mind. The annotated image shows the current planting patterns, calls out important questions about street trees and also points out opportunities for more tree planning.

Image: Sara Constantineau, 2021.

PHOTOGRAPHY IS GENERALLY THOUGHT OF AS A STAND-IN FOR SEEING

Photography is often expected to act as an equivalent to seeing and being in a place. The long-standing metaphor of photographs as windows onto the world perpetuates this conceptual position.[9] With this metaphor, the frame of the photograph is the frame of the window through which I view the landscape. *It's just like being there* – an oft-used phrase associated with photography. With their repeated use, these metaphors creep in and take hold, masquerading as the literal. The unexamined metaphor, or turn of phrase, slowly becomes treated as a fact or a given.

Wait, but doesn't photography show us what we see? Physiologically it does not. Most of us see with two lens-based structures (our eyes), not one. We have a cone of vision

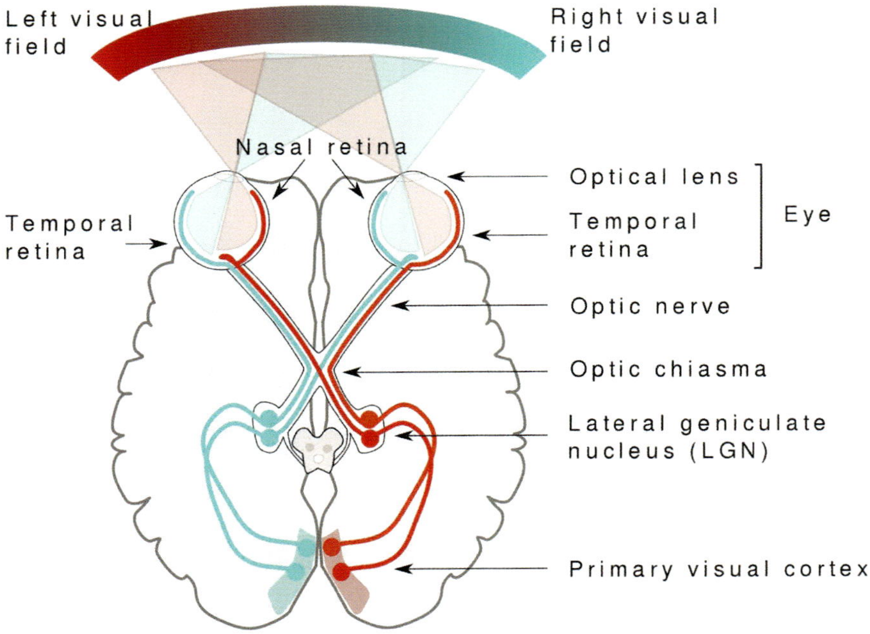

Figure 5
A complex set of neural pathways convey electrochemical signals from the retinas of the eyes to the visual cortex. The visual cortex is made up of multiple stages with specific cell types that interpret these signals. For example, "simple cells" are found in the V1 area of the visual cortex and "complex cells" are found in V1, V2 and V3 areas. Each of these cell types responds to different information. Once the visual cortex interprets this information, it is then sent on to other parts of the brain to be further analyzed by more specialized cells to assist in human functioning. See the National Institutes for Health "Visual Cortex" for several papers and detailed descriptions of the process of seeing.

Image: Miquel Perello Nieto, creative commons.

that is approximately 30–60 degrees in focus and 120 degrees of peripheral vision. We see in more of a circle of space, not a tight rectilinear frame. Our eyes (lenses) constantly move while we look at anything. Seeing is a complex brain process involving far more than lenses and a light-sensitive surface. Instead, the brain makes an image based on the interaction of multiple optical, physiological and neurological systems (see Figure 5).

At this point you might be saying, well, of course I know all of that. Indeed, we do know all of this, or at least most of it. Yet, it's easy to default to repeated conceptual constructions built around photography that started as metaphors and are now casually considered literal.[10] *Photography is a window onto the world. Photography is just like being there.*

This is when the concept of "photographic meaning," discussed in the following section, helps us understand the complexity of photographic representation (it's not like being there) and supports the practice of active landscape photography. Examining this relationship of meaning cultivates a more explicit understanding of the conceptual, psychological, emotional and cultural implications of making and using photographs – that there is much more to the process of photography than copying, or even recording. Approaching photography actively, as a complex form of representation, helps address in the multifaceted implications of how photography works.

THE RELATIONSHIP OF PHOTOGRAPHIC MEANING

Making and viewing photography is both a relational and a relative activity. Though photographs are casually considered static objects (as discussed earlier and in *ALP 1*), the *process* of making and looking at photographs is complex and multidirectional. Several theoretical constructs discussing this complexity, such as Roland Barthes' *Operator, Spectator, Spectrum*[11] and Terry Barrett's discussion of the relationship between "original context" and "external context,"[12] create precedents for the conceptual structure presented here. For this discussion, this structure is situated within the everyday uses of photography for landscape architecture and environmental design.

> First, we will examine the relationship of looking that creates photographic meaning. It is made up of four indivisible parts:
>
> Photographer
> Photograph
> Viewer
> Context

Each of these parts is constantly changing and is changed by the condition of each of the other parts. Photographic meaning is never fixed because the conditions of each of the parts are never static.

The *photographer* brings their unique experiences in the world to the making of photographs. These experiences are influenced by multiple factors, from culture, language, physical ability, race and gender to education, interests, relationships, socioeconomic status and region of origin (to name but a few). Some of these factors change over time to one degree or another. So, for example, the photographer I was ten years ago is not the photographer I am today because I have accumulated life experiences that have shifted my outlook on the world.

In the making of specific photographs, the photographer carries motivations that determine what kinds of photographs they choose to make or not make. Motivations can be guided by larger ideals – for example, aesthetic preference – *wow, that's really beautiful*. For our purposes, these motivations are more specific stated sets of criteria or goals – investigating a site for design, researching a particular material use, documenting a community meeting. In both cases, these motivations are usually underwritten by a combination of acknowledged biases, and unconscious biases created by the life experience of the photographer. For example, what one considers beautiful is a construct of past experiences built by what has been designated as beautiful by others. All of this comes together to influence how the photographer makes a photograph.

The *photograph* itself is a representation of the subject matter the photographer deemed important. The photograph is more than a container or box into which the subject matter is placed. Instead, the photograph is a tool of representation that influences how the subject matter appears. Through framing, point of view, time of day, composition, depth of field and exposure, the photographer uses the specific representational qualities of photography and the unique attributes of their camera to make the photograph of the subject matter.

As discussed earlier, the photograph is not a 1 to 1 copy of the subject matter, but a translation and interpretation of the subject matter through the representational qualities of photography utilized and influenced by the unique experiences and outlook of the photographer.

The *viewer*, like the photographer, brings his own life experience and thus biases to the viewing of the photograph. His life experiences, the cultural constructs that form his past and the ones he lives in now influence how the viewer looks at and interprets the information presented in the photograph by the photographer. While the photographer influences how the viewer sees the subject matter, the viewer also sees the subject matter in terms of his own conceptual and perceptual lenses. His background may assist him in, for instance, simultaneously accurately typing each of the tree species in the photograph, while also identifying a particular cultural event occurring. These same life experiences may not help him understand unique interactions happening between individuals or identify the historic periods of the architecture (see Figure 6). While someone else looking at the same photograph may name and interpret the unique architectural forms and be able to tell you about something that is not in the photograph. Such as the absence of the Erie Canal, which used to run through this very square.

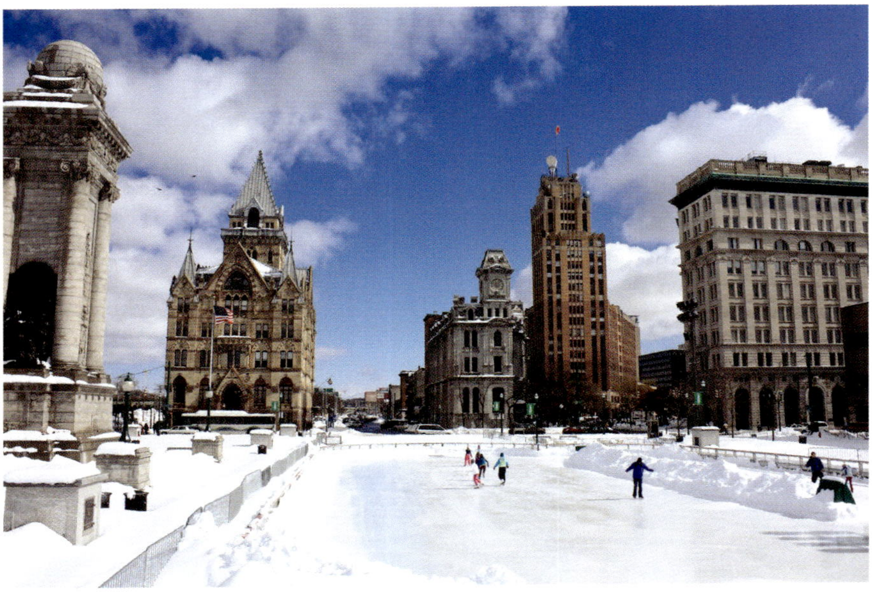

Figure 6
Each winter, Clinton Square is turned into a public skating rink. This image, looking to the east, shows the historic buildings that flanked the Erie Canal when it ran through the length of Syracuse, New York.

Image: Gavin Schaefer, February 2010, creative commons.

The *context* in which the photograph is viewed also influences how the viewer interprets and understands the subject matter presented in the photograph by the photographer. Another photograph of this same area in Syracuse, New York, made at a much earlier time might be viewed in an archive, cataloged with other similar photographs depicting historic events occurring in the square (see Figure 7). The same photograph, or a similar one made at a later date, could appear in an exhibit about what has changed in Syracuse over the last 120 years. Or the exhibit could be specifically about the Erie Canal. The photograph could be pinned up with other photographs showing different uses in the square as a way to help a landscape architect understand what uses occur and what changes could be made in the square to better accommodate those needs. The photograph might be viewed on a website promoting a particular event. It might be displayed as part of an email in which community members are invited to discuss how to overcome funding gaps for initiatives laid out in the Syracuse comprehensive plan.

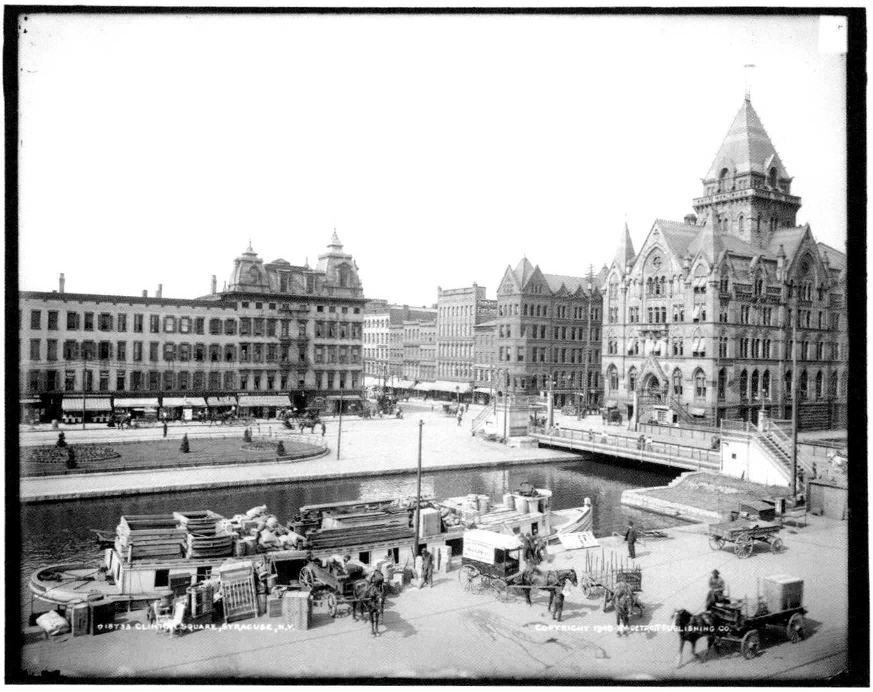

Figure 7
In 1900, the Erie Canal ran through Syracuse, creating a commerce hub around what is now Clinton Square. Today, those commerce functions have all but disappeared from this part of the city. Notice how this caption is working: a context for viewing is created by presenting specific information about the past with a discussion of current conditions.

Image: Detroit Publishing Company, Library of Congress.

Each of the contexts in which the photograph is viewed places it in a particular conceptual framework. These different situations color the way a viewer looks at and considers what is presented in the photograph.

Looking at the photograph in a historic archive, I may compare certain architectural details in the main photograph with other photographs. In an exhibit, I might be fascinated by looking at the clothing people are wearing, the abundance of shop signs (which no longer exist) and the arrangement of cars (see Figure 8). When shown this photograph in an email, I might think about how the past uses may or may not really fit what I think would be a better use for the future of this place.

As landscape architects and environmental designers, we always look at photographs in particular contexts. We gather photographs together from multiple sources to help us understand a place and make decisions for it. We view photographs on screens,

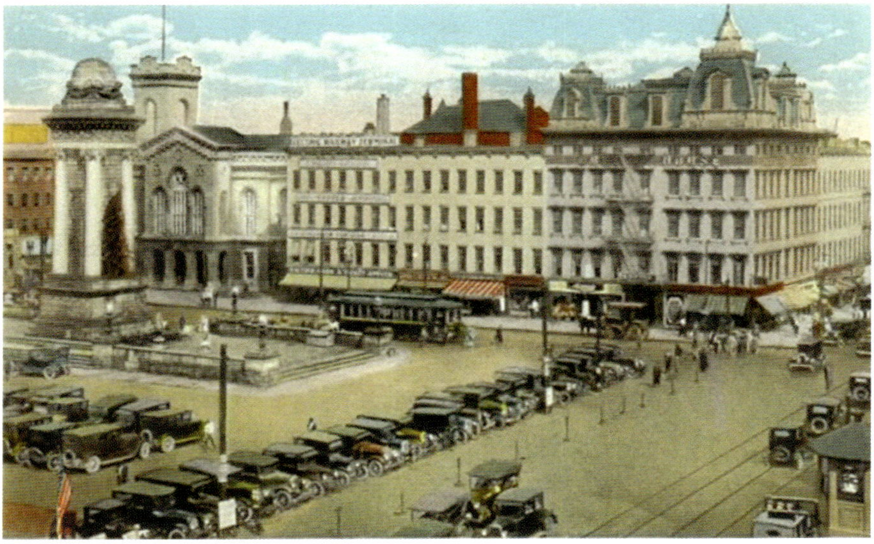

Figure 8
This 1920 postcard, made from a photograph, shows how the use of this area changed once the Erie Canal was moved in 1917. Cars are parked where the Canal used to flow. This view to the north looks at the Soldiers and Sailors Monument, dedicated in 1911.

Image: Unknown source, public domain.

pinned up, placed on a table, presented in a document or slide show presentation. And photographs are presented to us by clients, community members, stakeholders and investors. Not only are these photographs placed in physical contexts for viewing, they also are placed in conceptual contexts and contexts that weigh or favor certain values.

PHOTOGRAPHIC MEANING IS RELATIVE

One condition of the relationship between photographer, photograph, viewer and context is that a photograph can never have a single meaning. Another condition is a photograph does not mean or speak all by itself. The viewer brings her own life experiences, education, and biases to the interpretation of the photograph. And though the photographer influences how the viewer sees the subject matter, the photographer cannot maintain a kind of singular control over how a viewer thinks about that subject matter.

This brings into question another casual idea, that photography is universally understood, or has a kind of universal language. This idea has less hold than it used to in the mid-last century, especially as postmodernism[13] embraced modes of thought that valued the diverse and unique experiences of the individual viewer in the interpretation of any form of representation (in its broadest sense). As landscape architects, this condition reminds us that the diversity of people we work with on any project will have diverse understandings of the photographs presented as part of the design process.

Finally, this condition of relativity demonstrates how photography is not simply a visual portal opened through which we view a part of the world (see note 9). We cannot treat a photograph as a 1 to 1 stand-in for visiting the place. As environmental designers, we conceptually understand that, but on a daily basis, it is easy to default to treating photography as eyesight (site). Here, this book presents an array of methods to train and remind the photographer/designer to claim their active role in the making of photographic representations.

USING PREDETERMINED METHODS FOR PHOTOGRAPHY

At its most essential, using predetermined methods for photography encourages photographing a landscape beyond the photographer/designer's default set of passive moves and motivations. Applying methods to photography for site research and design instigates a conscious application of repeatable processes that guide and focus site investigation. Photographic skills increase over time as the methods are practiced and repeated (similar to photographing in a manual mode with a single lens reflex camera, as discussed in the following section). Practicing the methods turns passive photography into active photography. Over time, the photographer/designer cultivates new knowledge about the process of photography itself.

Many disciplines within environmental design use particular predetermined methods of investigation, data collection and knowledge creation. In order to engage in these methods, one must learn and practice the basics, or best practices. For example, I teach at a college of environmental science. Many of my students participate in lab work as part of their core curriculum. Before they are able to conduct studies or perform experiments, they must learn basics, such as safe techniques in the lab. They must learn methods, such as how to measure accurately with specific tools. These methods are practiced, and often the students must prove they can execute these core methods properly before they are allowed to conduct more complex experiments or be accepted into a faculty person's research lab.

At first, learning how to measure accurately, in any of our fields, can be challenging, from learning how to use architectural and engineering scales to learning how to survey. Attention must be paid and sustained. How must the pipette or beaker be held? How must the measuring tape be held and manipulated? Which mark do I use to start the measurement? Certain subtle movements are learned, often demonstrated by someone who does this kind of thing all the time. That person often makes the technique look easy. That person has internalized the techniques through repetition.

A former colleague of mine, architect Efthimios Maniatis, gives a demonstration to new design students about how to properly use a measuring tape. As a practicing architect, Maniatis has measured innumerable interior and exterior spaces. When I told the students that someone is going to come in and teach them how to measure with a tape, the students thought at first, *well sure, I can use a measuring tape, who can't?* Anyone can go buy a measuring tape at the hardware store. A measuring tape does not come with instructions. Everyone has a measuring tape in their junk drawer at home.

Maniatis comes in and hands out tapes with metal blades – 25′ and 50′ tapes. He asks the students to measure some things in the classroom. First it is easy, such as the length of the wall at the floor or the surface of a tabletop. Then, Maniatis asks them to measure the height of the wall, floor to ceiling. This is when the chaos starts. As the students attempt to extend the tapes longer than a couple of feet, they clank and bang as they lose tension and bend over. Maniatis asks them to measure a complex corner, including a structural column. He asks them to measure the inside of a box by reaching into it with the tape. He then asks them to try to make these measurements without a partner, just on their own. Then the tapes really begin to bend and crack and clank. The students start to realize there is more to this than they thought.

Then Maniatis tells them all to sit, put their tapes down, and watch. Deftly, he demonstrates how to walk the metal measuring tape blade up a wall and how to access a hard corner. He shows them how to read the tape at the curve it makes when bent into a corner. Then he talks about the common marks on the case and blade and how to use them. Finally, he discusses the tolerance on the hook, what that tolerance is for and why the hook moves.

During his demonstration, the tape measure does not lose tension and unintentionally bend or make that cracking noise a single time. Maniatis makes it look easy. He shows the students it can be easy if you learn the basic methods of manipulating the measuring tape. And then practice.

I hope the metaphor here is obvious.

This book provides a similar starting point, presenting a set of fundamentals that help the photographer become more rigorous in their photographic process, through training in the basics.

Here is another metaphor that can help us think about the utility of learning and practicing predetermined methods for photography:

What if we were to think of a camera like a microscope? Microscopes can be quite simple – a set of lenses that magnify something. They can also be extremely complicated, utilizing computer programs to process and synthesize information. Most microscopes, like contemporary digital cameras, are reliant on both lenses to assist in seeing and complex electronics, sensors and programming to enhance what those lenses see.

Yet, we don't think of microscopes as telling "us what is there" nor do we think they do the work for us. Instead, we think of microscopes as tools that are used as part of the method of studying of whatever we put in the microscope. Microscopes help us see things in ways we are not able to with our naked eyes.

We engage in a set of methods for using microscopes. And we have a plan to look for something in particular when we use them. There are a set of procedures utilized in the use of a microscope. There is another set of more specific procedures based on what is placed under or in the microscope. And we don't use a microscope while simultaneously doing other things.

Though certainly not perfect, I hope this metaphor instigates a different attitude toward our use of photography. It gets at considering photography as something we thoughtfully utilize in a methodical manner as a form of research and investigation.

PRACTICING METHODS, METHODS AS PRACTICE

As introduced previously, photographing with methods takes practice. We can practice like any other procedure we use as part of the design process, from surveying, to making 3D models, to presenting to a community. At first, the methods may seem clunky or even artificial, yet over time, the process of the methods becomes internalized, requiring less time to think about them. This is similar to photographing in a manual mode with a single reflex camera. One must, for every photograph, meter the light and choose the setting for both the aperture (f-stop) and shutter speed. This determines the exposure of the image, while also simultaneously influencing the depth of field. The reciprocal relationship between the two is always negotiated by the photographer.

This process, or the method of determining the relationship between the exposure and the depth of field, is a physical process for the photographer – they manipulate the controls of the camera to set the f-stop and shutter speed. At first, this takes time and feels very awkward. I must simultaneously remember the rules about aperture and shutter speed while remembering what the light meter readings mean while considering how I would like my subject to appear with both exposure and depth of field. I must rotate physical controls on the camera to set things up and pay attention to how that changes the reading of the light meter. And then consult what I have learned about how this setting will influence how my subject looks. And finally, I must make the decision if this is what I would like to do. With practice, the photographer is able to very quickly both make a choice about the appearance of the photograph and negotiate the controls of the camera to set aperture and shutter speed to yield the desired result.

This process of determining exposure and depth of field can feel daunting, and it seems like it takes a good deal of time, especially at first. It can be frustrating and confusing. Many people give up. Yet this process is essential to learning how to make photographs and be in control of the appearance of the photograph.[14] If I were to skip this process by sticking to "auto" mode because it's "easy" it would mean I don't really know how my camera works and I don't really choose how my images appear.

After practice, making these choices while photographing occurs quickly and much more fluidly. My fingers easily work the controls because they've done it countless times. When I get a new camera, with slightly different controls, it takes a bit to adjust, but I know it must be part of the process. This may feel similar to the process of learning a piece of software's key commands or getting good at Photoshop or learning all of the updates on a piece of GIS software. It may be similar to the process of learning new building codes and then actualizing them through the redlining process.

With practice, the reader will feel the same way about many of the methods presented here. At first, they may seem stiff or artificial, but by practicing and learning from the results of working with the methods, the benefits of the process of the methods become clear, and the negotiation and execution of the methods become internalized. Once the methods become more comfortable, they help the photographer become more active and conscious while photographing. Stronger engagement is cultivated during the full process of photographing, and it becomes a form of focused investigation.

MORE THAN THE SUM OF ITS PARTS

Conceptually, photography is incredibly complex. Now, in the early part of the 21st century, nearly 200 years since the invention of photography, a significant body of writing acknowledges and demonstrates how photography is not a thing, with one result, but many things that change over time. A photograph has a life influenced by the motivations of its original making, the contexts in which it exists and the background of the viewer looking at it. The novelty of earlier conceptual observations, such as that photography is magic,[15] removes the notion of originality[16] and functions with a universally understood visual language,[17] have all but worn off now after late 20th and early 21st century photographers and theorist have scratched at the surface long enough to better understand how photography works.

Here, we scratch at that surface for the sake of taking hold of this ubiquitous form of representation in environmental design. Though our metaphors thus far have been primarily interested in the act of measuring or observing an actual thing, we can start to understand that photography represents more than the actual.

A way into this concept is through noticing how the process of photography is far more than the sum of its parts. "Parts" here refer to the methods utilized, the site visits made, the people making the photographs, the camera itself, the subject matter photographed and the resulting photographs.

Photography is a catalyst for synthesizing complex knowledge and experience in order to confirm or create new understandings about a place. Photography motivates a dialogue about a place. Take, for example, one of the core methods presented in Part I: Methods, which includes consciously asking the question, "Why am I photographing this?"

The answer can vary widely. Yet it is the posing of the question in a deliberate way for each photograph made that triggers active acknowledgment and synthesis. This is where photography becomes more than the sum of its parts – the synthesis is something more than each piece or part of the process.

Photographs are made to show a particular interest in something. That "something" is often related to a larger phenomenon or system. For example, the photograph of a stream bed presents an instance of stream degradation due to an increasingly flashy storm system. The photograph focuses of the degraded condition of the stream edge, not the rest of the nested interrelationships within the system (see Figure 9). Yet, this photograph was made with full consciousness of characteristics within the larger system. This watershed's "flashiness" is exacerbated by an increase in impermeable surfaces and climate change

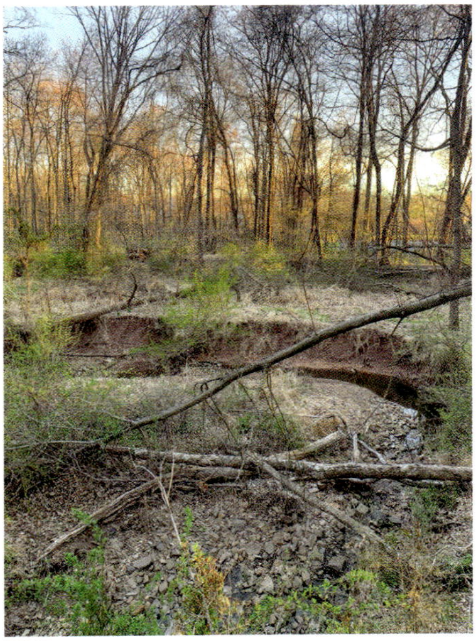

Figure 9
The sharply eroded edges of this stream indicate a repeated occurrence of high-water volumes flowing in fast bursts. This erosion is exacerbated by increased impermeable surfaces within the local watershed. In the distance, between the trees, one can barely see the hard surface of an adjacent parking lot. Bridgewater, New Jersey.

Image: Sara Constantineau, 2021.

intensified storms. The act of photographing this specific part of the watershed synthesizes complex knowledge about the system. It becomes about more than photographing the actual.

Another example may be that I am photographing a historic square somewhere in the American South. My photographs of it may show many people sitting on benches in the shade of large oaks or sycamores on a Sunday afternoon after church. My photographs show the activity of the present moment of the photograph. Yet I may also know and become more fully aware, through the act of photographing, that my photographs show a change. I know through my research that this square used to be a site of a confederate monument. I may never have witnessed it in that state since I am visiting for the first time several years after the monument's removal. The process of my photography synthesizes the past and the present. I may wonder at the extraordinary changes, physical and societal, that have occurred over the last several years. I think about what is to come or what may not have happened yet. My photographing acts as a touchstone to this unfolding history.

Showing up to a place and photographing becomes a complex convergence of memory, experience, learning and projection into the future. This disrupts the idea of photographs as inert or silent things. Instead, the photographer (and the viewer) can experience a swirling intermingling of phenomena, seen and unseen, when we understand that photography is more than the sum of its parts.

Through practice and active engagement in photographic methods, photography becomes a more thoughtful process of representation. Instead of producing an abundance of throwaway images (which I must search through at a later point to find something helpful), photography can be an action of synthesis and knowledge creation in the moment of making the photograph. Through practice, photography becomes an active investigative tool. Making photographs on-site becomes a primary form of research. Utilizing predetermined methods is the means through which to achieve this engaged state.

HOW TO USE THIS BOOK

The methods presented here teach the reader how to practice active photography. The methods present multiple lessons that help the photographer increase awareness while photographing. How photographs function as complex forms of representation is also discussed throughout. Each method addresses common scenarios in the everyday practice of landscape architecture and environmental design.

Part I: Methods presents a primary set of basic skills that can be used during most photography sessions. Each helps the photographer more actively engage with the process of photographing by increasing attention through specific prompts. These are meant to get the photographer out of a passive mode of pointing and shooting. As these methods progress, they become more layered in their steps. Part I: Methods provides specific training in photography for landscape architecture and environmental design disciplines, filling the existing gap.

Part II: Practices present more complex methods inspired by specific contemporary photographers' projects and practices. The underlying concepts of these photographers' work are discussed and then extrapolated into situations encountered in landscape architecture and environmental design.

THIS BOOK IS WRITTEN FOR SEVERAL AUDIENCES

First, these Methods and Practices are presented to be easily assessed and quickly implemented by practitioners, students, researchers, and educators. For example, a practitioner or an individual student could take themselves through the Methods in Part I as a way to hone their photography practice. Specific Methods could be utilized by a design or research team during a site visit. While Part II: Practices serve as case studies to be singled out and applied to specific projects based on specific needs.

Second, this book is written and structured to act as the core text for a photography course within an environmental design curriculum, thus filling the current gap. These methods can also be embedded into a studio course. For instance, a studio could utilize Part I: Methods as a guide for engaging in site investigation. As case studies, Part II: Practices can be applied or extrapolated for design projects, research, capstones, or theses. An advanced studio, practicum, field study or upper division course could choose a subset from the Methods and the Practices to address the needs of a specific project.

This book is an outgrowth of a set of courses I have designed and taught over the last decade and a half, starting in landscape architecture and then extending into many of the landscape-interested fields. Students within landscape architecture, as well as other environmental design disciplines, from architecture and planning, to environmental forest biology and restoration ecology have found these courses to be applicable to their daily practices. For example, one student shared in an end of class survey:

> As a conservation biology major (EFB), this course was very different from all my other courses. . . . Photography can be extremely important and useful as a skill for EFB majors as it can be used in presentations, be the difference between successful and unsuccessful public awareness campaigns, and be used while doing academic research. I think this would be valuable for students in any [environmental science] major to experience at some point during their undergraduate life.

As time has passed, and I have worked with a broader spectrum of students from various environmental science disciplines I have noticed how students within this larger group very positively respond to both the core Methods found in Part I and what they might characterize as the "more experimental" Practices in Part II. Many of the approaches we take for granted in landscape architecture, based on making things and experimenting through design, are welcome challenges to students in the environmental sciences who are looking to stretch themselves and vary their skill sets. (I will point out that the practices of environmental science are forms of design in that their work both influences the understanding of and explicitly changes the landscape.)

These Methods may also be consulted to get unstuck during any research and design processes. They are prompts for thinking differently. They disrupt unconscious working practices. They encourage the photographer/designer to slow down and take time to pay attention.

There is always time to learn how to photograph in a thoughtful and methodological manner. Time in photography is measured in increments of a second. We think as fast as the shutter releases. Each of these predetermined methods asks for seconds of consideration, one second at a time. It only takes a moment to initiate active photography.

Let's start.

NOTES

1 See Anne C Godfrey, "Do We See What We See? More Than Icons for Landscape Photography," in *Active Landscape Photography: Theoretical Groundwork for Landscape Architecture* (London: Routledge, 2020), 61–69.
2 For example, one early advertisement reads: "Load in Daylight with our six exposure film cartridges and are *[sic]* so simple they can be easily operated by any school boy or girl." Eastman Kodak Company, *Kodak Advertisement for the Brownie Camera*, 1900, Print, 4 1/2 × 6 inches (Rochester, NY: Eastman Museum).
3 Peter Buse, *The Camera Does the Rest: How Polaroid Changed Photography* (Chicago, IL: University of Chicago Press, 2016), 8–12.
4 Polaroid Company advertisement, 1978.
5 Samsung, "How to Use the Camera on a Galaxy Smartphone," accessed August 27, 2021, www.samsung.com/uk/support/mobile-devices/how-to-use-the-camera-on-a-galaxy-smartphone/.
6 Anne Godfrey, "Tanner Springs Moments," *Landscape Architecture Magazine*, February 2020, 110.
7 Anne C Godfrey, "Making Photographs: Making Landscapes," in *Active Landscape Photography: Theoretical Groundwork for Landscape Architecture* (London: Routledge, 2020), 33–47.
8 Also see Anne C Godfrey, "Landscape Architecture and Photography: A Ubiquitous Relationship," in *Active Landscape Photography: Theoretical Groundwork for Landscape Architecture* (London: Routledge, 2020), 7–15.
9 For example, in the most sublet of ways, photography is used to describe site, while not acknowledging that it is a photograph that presents the visual representation of the site. It is as if the photograph does not exist when it is used to show a place. Respected anthropologist Setha Low engages in this practice in her seminal book *Spatializing Culture*. Photos are used throughout to discuss, primarily, how people utilize urban spaces. Photography as a concept and term are not called out. The word does not appear in the index. The act of framing and how that creates a sense of space is also not discussed. This is no specific fault of Low's. *Spatializing Culture* has proved to be an influential work not only in anthropology but also across fields interested in humans in space, including landscape architecture. Instead, this is an example of how photography is "used" without acknowledging how it works and influences the appearance and thus perception of subject matter. See Setha Low, *Spatializing Culture: The Ethnography of Space and Place* (London: Routledge, 2017).
10 See George Lakoff and Mark Johnson, *Metaphors We Live By* (Chicago, IL: University of Chicago Press, 2011).
11 Roland Barthes, *Camera Lucida: Reflections on Photography* (New York: Hill and Wang, 2010), 9.
12 Terry Barrett, *Criticizing Photographs: An Introduction to Understanding Images* (New York: McGraw-Hill, 2012).
13 The term "postmodernism" is used broadly here to create a loose multisided, ever-changing container for the set of critical theories that started to develop during the latter half of the 20th century, including postcolonial studies, feminist studies, critical cultural theory, poststructuralism, affect theory, queer theory, Anthropocene studies and critical race theory (to name a few).
14 Landscape photographer Richard Misrach comments, "Learn the basics of how to expose develop and print. Stick to one project, one camera, one lens. They will become natural to you." Richard Misrach, *Richard Misrach on Landscape and Meaning* (New York: Aperture, 2020), 18.
15 Warner presents the many early instances of photography being compared to magic in her volume: Mary Warner Marien, *Photography and Its Critics: A Cultural History, 1839–1900* (Cambridge, England: Cambridge University Press, 1997). Talbot, so taken with his invention, considers it to be like magic and refers to it as so in his writings. William H. F. Talbot, *Some Account of the Art of Photogenic Drawing: Or the Process by Which Natural Objects May Be Made to Delineate Themselves Without the Aid of the Artist's Pencil* (London: Printed by R. and J. E. Taylor, 1839).
16 See Walter Benjamin, *One-Way Street and Other Writings* (London, England: Penguin, 2009).
17 Such as the exhibit: The Family of Man, New York, NY, Museum of Modern Art, January 24-May 8, 1955.

Terms, Assumptions and Concepts

Before moving forward, this section introduces and discusses some reoccurring terms and assumptions. These terms and assumptions create a background for the methods and articulate some of the ground rules for underlying concepts. Slippage will occur. Debates about the contemporary use and applications of some of these terms and assumptions continue to evolve. Here I have tried to combine broader definitions, considering the audience intended for this book, with a brief introduction to some of the contemporary discussions currently in play.

Richard Shiff, in his essay on the painter Joan Mitchell, states, "Words are signs that refer by generalizing, and imprecision accompanies their use."[1] This frame or approach Shiff articulates is informed by general concepts from semiotics,[2] which acknowledges how the meanings of words shift based on context. As we discuss utilizing photographs for design decision making, I will provide some definitions for these terms to place them within this specific context. Some of these terms are under erasure and will continue to be such, so I provide some background discussion to avoid misunderstandings. Words can be imprecise tools we use to describe precise things. And, like photographs, the meaning of words is not entirely fixed.

SITE

The term site is used throughout. For our interests, site is defined broadly. A site may be the following:

- The clearly delineated extent of any design project (often contractually defined). This will often be referred to as the "project site" or "the design site."
- A subsite within a project site.
- The context of the project site. This context may be immediately adjacent or much larger in extent.
- A neighborhood, district or other defined subarea in a village, town or city.
- A region, watershed, or other geographically described space.
- A cultural landscape.
- A place defined by its ecology, such as a salt marsh or oak savanna.
- A bioregion.
- A subsite within any of these larger definitions of site.

DOI: 10.4324/9780429284649-2

Site refers to any and all aspects, characteristics, conditions, histories and phenomena of the place. These may be surface, visual and tangible. They may also be subsurface, invisible, or function at a larger or smaller scale than normal human preception. For example, climate or hydrology are characteristics of a site. So are human demographics such as age and income level. Photography aids in the representation, denotation, and synthesis of any of these characteristics.

SITE PHOTOGRAPHY IS UNIQUE

The type of photography discussed here is site photography, which is a particular kind of photography. Site photography is investigative and made to assist in the research and design process. This is in contrast to, for example, personal photography for social media, which often works in quite different ways. There could be an overlap between the two types, but functionally, motivations behind these two types of photography often differ.

This is a great example of the slippery situation that exists when we attempt to neatly define and categorize things. I bring this up to highlight the condition that all photographs do not do the same thing – all photography is not the same.

For instance, the goals set out for site photography (with the implementation of a set of predetermined methods) are different than the goals set out for built work photography.[3] Neither type of photography is better than the other, they just work differently and need to be understood under different criteria. A simple example of this is that site photography is used to figure something out, not show something off. While built work photography uses aesthetic conventions of "beauty" or "unity" (for example) to show the success of a finished design project.

PHONES AND CAMERAS

The assumption is that we always have a camera available because having a mobile phone means having a camera. This condition makes photography very accessible and has exponentially increased the number of photographs made on a daily basis.[4] Similarly, photos are shared constantly both publicly and privately through social media platforms, text messages, and via email and other direct to individual internet-dependent communication applications.

Photography is used to communicate in so many different ways – location, interest, event, emotion, note-taking. *Hey, I'm here, and isn't this amazing!* (See Figure 1) Photography is also used as a stand-in for memory, in a very utilitarian way. *Let me photograph this menu posted outside this restaurant so I will remember to order takeout from them later this week.* Or as a note-taking device. *Let me photograph these luggage tags in case I lose them* (see Figure 2). Or as a way to cheer someone up (see Figure 3). This is just a small handful of examples.

The point is we use photography to communicate in all kinds of very flexible and fluid ways. Taking a photograph is very easy, with no real expenditure of time or resources necessary. This is similar to what graphic designer Aaron Draplin (creator of Field Notes) frequently says, "Pixels are free!"[5] The same is true for photography. The only thing we run out of is space – as in storage and memory.

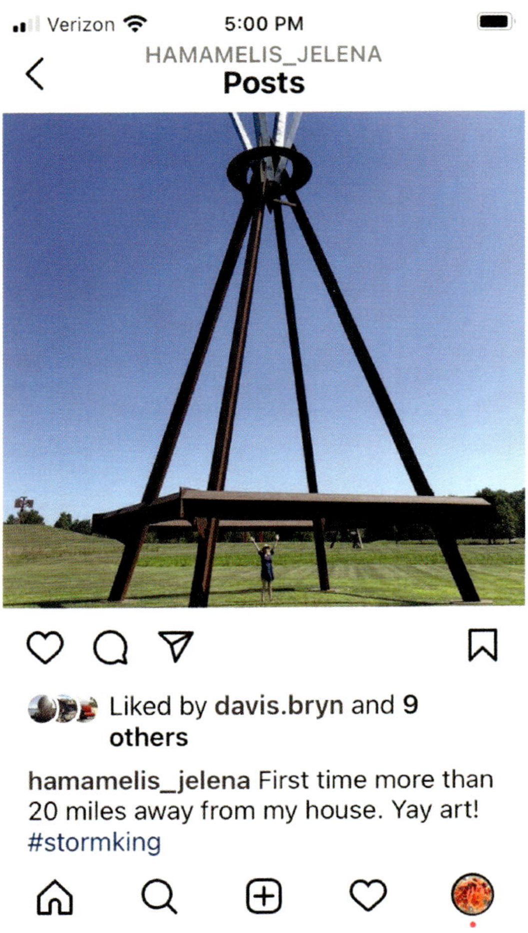

Figure 1
An Instagram post shares how excited I am to travel again after the first phase of the COVID-19 pandemic ebbed. Here, I share a photo of myself under a Mark di Suvero sculpture at Storm King Art Center in the Hudson River Valley of New York State. This type of photography is intentionally made to share with others. In this case, the photograph is communicating about a place and an emotion during a very specific time period.

Image: Don Zolidis, June 2020.

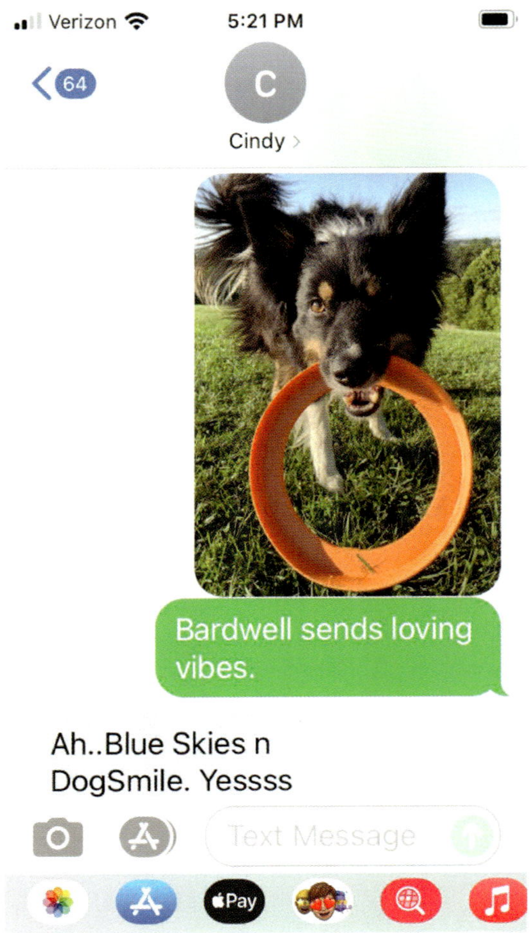

Figure 2
Here I send a cheerful photograph of Bardwell the border collie to a friend in Oregon. The message was sent in hopes of lifting her spirits during the devastating fires that swept across the state in September 2020 and sent the air quality index into a dangerous deep purple alert with readings between 350 and 515.

Image: Anne C Godfrey, September 2020.

How do these conditions affect our daily use of photography during the design process?

1) We can always make a photograph. The pro of this is that we might find ourselves in a unique situation or circumstance and photograph something that acts as inspiration, serves as evidence, or helps us understand something. And we can photograph it without thinking twice. This gets to the adage that the best camera you have is the one that is with you. The con is we might not work as hard to really look at and examine a particular phenomenon because *hey, I have my camera with me! Snap!* or rather *Boop!* It's easy to take photographs. I can photograph without thinking twice.[6]

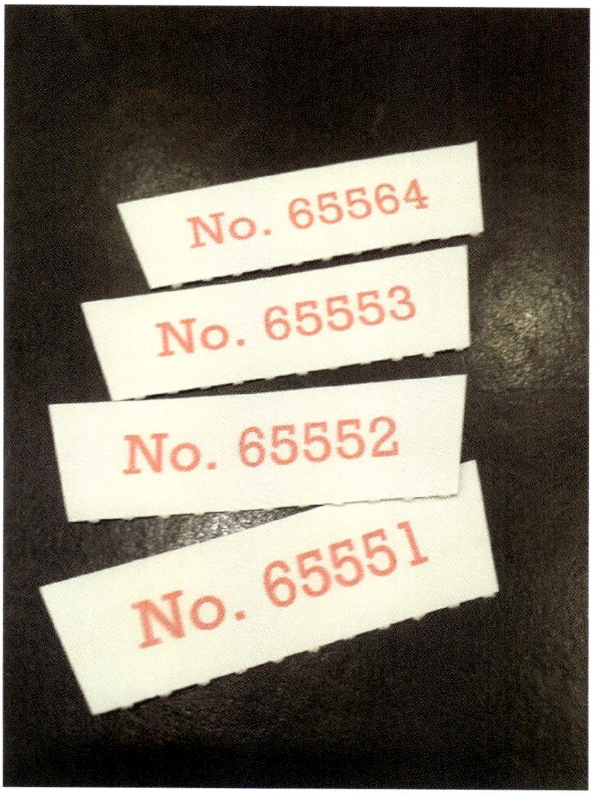

Figure 3
Photographs of luggage check tags from the Graduate Hotel in Minneapolis are made to both remember the numbers and safeguard them if they get lost during a busy day. Photography is used as an assistant to memory and as a form of note-taking.

Image: Anne C Godfrey, March 2020.

2) We can make as many photographs as we want. The pro of this is photography can be a way to take notes. I don't have to worry about wasting a precious frame of film to photograph a list, or a book cover, or a silly thing that made me laugh or as many pictures of my dog as I want. I can photograph every interesting plant I see. I can photograph every construction detail. I can experiment with making multiple photo collages in place. I can mess around with some exposures. I can make a lot of mistakes to figure something out. The con is having this ability to make photographs of whatever I want, whenever I want, helps me stay in the mode of thinking "if I take enough photographs then one of them will turn out." This leads to taking a whole bunch of photographs, but hardly looking at any of them. I hardly look while I am taking the photograph, and I hardly look later as they stack up in my digital album and I flick through them. Photography becomes more about chance, instead of an intentional act.
3) I can share photographs whenever I want. The pro is this can be very helpful for a design team with members in different locations. Photographs for decision making can be instantly shared across the planet. I can ask someone to show

me something in real time. Or I can ask someone to find a particular photograph and share it with me instantly. Or someone can photograph a sketch and share it right away. The con is we might get overly caught up in this sharing. We might be more involved in our "sharing" than experiencing the actual place we are sharing about.[7] This gets back to not being present and experiencing the moment because the act of photographing-and-sharing dominates our attention. The ease with which we can photograph anything anywhere and share it anytime is a condition of mediation. Engagement in the constant action of photographing-and-sharing mediates our experience of the world.

SUBJECT AND SUBJECT MATTER

Like "site," the definition of the term "subject" is treated broadly. For this work, subject is anything that is photographed. This can be from a small detail, to a person, to a large landscape. "Subject matter" is used interchangeably with "subject" in this book. "Characteristic" is another word closely associated with subject. For example, *The subject matter of the photograph shows many different landscape characteristics because the photographer is standing at a distance.*

Subject is a fraught word. In some instances, the use of the word "subject" becomes a term and concept put to the task of objectifying. In other situations, the idea of "subject" works toward abstracting the specifics of a unique individual or characteristic. Sometimes the use of the word "subject" is deployed to distance the viewer, examiner or person making a judgment from the specifics of an individual. This also expresses how the word "subject" can be used to abstract, and essentially erase, unique qualities. These uses of the word "subject" are not the intended uses for this work.

Here the use of the words "subject" and "subject matter" are presented in a neutral way – or perhaps a better word would be *expansive*. "Subject" and "subject matter" refer to "that which appears within" the representation of the photograph. This becomes a broad consideration. Subject is *everything* within the photograph, not only the primary subject that creates a focal point. This is important for landscape architecture in that the landscape is not just scenery, providing a background or a foil to a subject. Everything that appears in the photograph – in the fore, mid and background, in focus or out of focus, large or small – is the subject matter.

DESIGN

The word "design" is used to describe how humans alter the landscape. This can be obvious, such as a design for a public park, to more subtle, such as the design of a watershed implemented through planning and management strategies. Any set of planned techniques or interventions utilized by humans to change landscape conditions is a form of design.

Thus, this paring of terms appears throughout the book: "landscape architecture and environmental design." Environmental design refers to all of the disciplines that take part in the planning and altering of the landscape. This includes not only landscape architecture, architecture, planning and engineering but also many of the disciplines in environmental science that work directly with landscape systems. This is a broad definition of design. Yet it directly acknowledges that humans intentionally alter landscapes in many ways.

Photography plays a crucial role in design decision making because it is used to investigate, document, assess, understand and synthesize characteristics in the landscape.

RESONATE AND RELATABLE

Many of these methods have been developed through my work with students in my standalone courses on photography. They have also been inspired by my ongoing conversations with practitioners as we discuss both specific projects and general ideas about the use of photography in design.

This results in a qualitative approach presenting methods that seek to resonate with the users. Resonating is as simple as, *hey I tried that, and I did notice a difference in how I photographed*. The idea is that the designer/photographer becomes more engaged and is able to pay closer attention during the process of photographing.

The methods are relatable because they are easy to pick up and do. There are no barriers, like specialized equipment or training, to engaging in these methods. Of course, highly specific advanced photographic methods or technologies are utilized for certain projects. Some examples of this kind of work are presented in the next book in this series *Active Landscape Photography: Diverse Practices*. Yet, basic everyday photography often forms the photographic foundation for our design investigations and decision making process. The methods here are crafted so that anyone in an educational or professional setting can pick them up and immediately start using them.

This collection of methods is presented as a place to start. They create a set of explicit steps for practicing active photography, something currently missing in our environmental design fields. They also present a foundation from which anyone can create their own methods, extrapolating upon what is presented here to make something new.

NOTES

1 Richard Schiff, "Add Red, or Not," in *Joan Mitchell*, eds. Sarah Roberts and Katy Siegel (New Haven: Yale University Press, 2021).
2 For example, see Ferdinand De Saussure, *Course in General Linguistics*, trans. Roy Harris (1916; Chicago: Open Court, 1998); Roland Barthes, *Elements of Semiology*, trans. by Annette Lavers and Colin Smith (1964; London: Jonathan Cape, 1967); Louise M. Rosenblatt, *Literature as Exploration* (1938; New York: The Modern Language Association of America, 1995).
3 As I discuss in ALP 1: Anne Godfrey, "Do We See What We See? More Than Icons for Landscape Photography," in *Active Landscape Photography: Theoretical Groundwork for Landscape Architecture* (London: Routledge, 2020), 61–69.
4 Marvin Heiferman, *Photography Changes Everything* (New York: Aperture, 2012); Peter Suciu, "A Photo Used to Be Worth a Thousand Words, but Thanks to Social Media Photos Have Lost Their Value," *Forbes*, October 24, 2019, www.forbes.com/sites/petersuciu/2019/10/24/a-photo-used-to-be-worth-a-thousand-words-but-thanks-to-social-media-photos-have-lost-their-value/?sh=66263cbf607f.

5 Aaron James Draplin, *Draplin Design Co.: Pretty Much Everything* (New York: Stewart, Tabori & Chang, 2016).
6 Julia Soares and Benjamin C. Storm, "Forget in a Flash: A Further Investigation of the Photo-Taking-Impairment Effect," *Journal of Applied Research in Memory and Cognition* 7, no. 1 (2018): 154–60, https://doi.org/10.1016/j.jarmac.2017.10.004.
7 Now a common discussion about photo sharing on social media. For example, see Cal Newport, *Digital Minimalism: On Living Better with Less Technology* (London: Penguin Business, 2020).

PART I

Methods

These methods introduce straightforward activities to heighten awareness during the photographing process. They get the photographer out of passive default modes of pointing and shooting and into active processes for making photographs. The prompts presented can be applied to most situations.

The first four methods create a core set of prompts to be used whenever one photographs. The successive methods introduce more layered and conceptually sophisticated approaches. All these methods are situated in everyday practices in landscape architecture and environmental design.

DOI: 10.4324/9780429284649-3

Method 1 Slowing Down

Helpful with: Activating close looking and seeing while photographing. Overcoming the tendency to think the camera sees for you.
Prep: Willingness to give this method a try.
Post-production steps: Slowing down when viewing the photographs.
Other materials: A timer or stopwatch.

DISCUSSION

Time and the activity of photographing are intertwined. Photography marks time. This chapter lays out straightforward and accessible methods for slowing down in order to pay closer attention while photographing.

What does "slowing down and paying attention" mean? Why is it helpful?

Photography is often treated as a speedy, quick endeavor. Digital photography has led to limitless photography – the click of the shutter, or push of a button or tap on the screen allows us to make as many photos as we want. The finite nature of film photography no longer exists in everyday photography. Now the only limits are battery life and storage capacity. Because of this, photography can be done almost thoughtlessly.

Using photography this way can become an unconscious replacement for seeing.[1] Digital photography saturates daily life through social media and any of the various photo-interested, internet-based platforms. Photographs are viewed all of the time, throughout the day, almost whenever a phone is picked up or any other kind of screen is used.[2] Photographer Catharine Opie observes the implication of this saturation, "We've forgotten what it is to really look, to be involved. We've become this image-based society where we are no longer looking."[3]

This points back to the condition discussed in the introduction: the camera is expected to do the looking in the moment – it is treated as a replacement for sight and believed to replicate what is seen with the eyes. Because we use photography in such an unconscious and untrained way, not regarding it as a skill to be learned and practiced, nor treating it as an active method for observation, this non-engaged use treats photography as a prosthetic memory, again taking photography for granted.[4]

Often, without consciously realizing it, we simply lift the camera up to take a picture, release the shutter, bring the camera down and then walk away. Often, not much time is spent looking at the landscape either before or after we take the photograph. It's easy to walk around a landscape site and snap, snap, snap. I have observed this countless times in various settings, such as students arriving at a design site for studio,

DOI: 10.4324/9780429284649-4

watching colleagues photograph for a project and as a voyeur watching people take photographs at National Parks or at significant sites of design, like Little Island in New York City[5] or Millennium Bridge in London.[6] In many of these cases, the people making the photographs are not actually looking closely at the landscape subject (see Figure 1).

Next time you are out photographing, watch what other people do when they are photographing. Pay attention to how long they look at the subject before and after they photograph it. It's interesting to spend some time at a viewpoint parking lot, such as in a National Park or Heritage site, and watch what people do. Often they park, get out of the car with the camera, walk to the best photo spot, bring up the camera, take the photograph and then almost immediately turn around and walk back to their car, or start talking with their companion or post the picture on social media (see Figure 2).

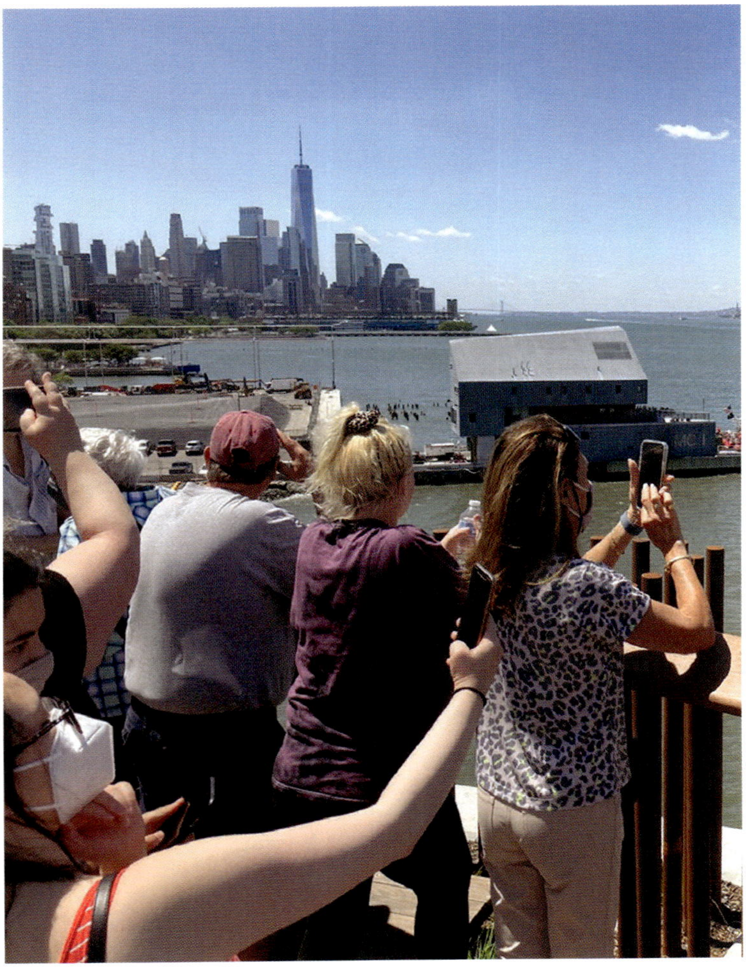

Figure 1
To manage the huge amount of interest in the opening of Little Island in May 2021, a timed ticket entry system was put into place – in part due to the ongoing spread of COVID-19. Here, at one of the elevated outlooks, tourists look south to the Lower Manhattan skyline and the Statue of Liberty. Notice the number of phones used to take selfies and photos of the view. I observed countless people traipse up to the top of the lookout, take some photos and selfies with their phones and then immediately turn around and walk back down.

Image: Anne C Godfrey, May 2021.

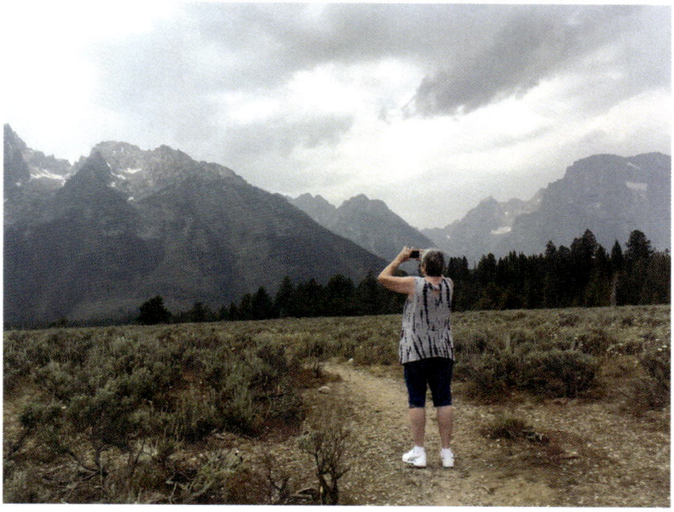

Figure 2
Grand Teton National Park is another well-known photo destination in the United States, made famous in part by Ansel Adams's photographs of the park. As seen here, tourists will drive into a turnout, get out of their car, take a photo of the view and then immediately turn around, get back in their car and drive away. This is a common way tourists take photographs in many of the National Parks around the United States.

Image: Anne C Godfrey, August 2018.

This is the most literal example of *taking* a photograph.[7] Walking up, taking a landscape scene with your phone and then walking away with it in your pocket. Not much time was invested in looking with the eyes, pondering, being in the place – perhaps less than a minute for the whole procedure. And the actual moment of taking the photograph – just a split second. This is pointing and shooting.

It is so easy to take, take, take. It's easy to forget to spend some time. None of us are immune to this activity. I still point and shoot when I am distracted or in a hurry.

An antidote to this passive way of using photography is to actively make a little bit more time for each photograph as a way to start really looking again. Setting aside a small amount of time is a way to reconnect to the idea that *photography is a process* and an opportunity to learn something.

THE METHOD: 5 SECOND RULE

It's easy to make a little bit of time to *look and see* with the camera. All it takes is 5 seconds.

The method is simple:

Before making a photograph, look at the screen or through the viewfinder for 5 seconds.[8]

Then release the shutter.

Try this right now. Set a timer for 5 seconds. Choose a place to look at either out the window or in the room you are in. Start the timer and look at the chosen place. Look for the whole 5 seconds.

What did you notice? How much did you sense your eyes moving? What did you think about?

Were you surprised by how long 5 seconds felt?

5 seconds is a nice amount of time. It is small temporally but can feel large experientially.

By delaying the release of the shutter, space is created to look, pause, see, think, reflect, cultivate awareness and remember. This transforms what is often a mechanical and automatic activity (taking) into one of contemplation and engagement (making).

5 seconds makes space for conscious looking and seeing and thinking. It also makes space for conscious feeling and noticing, such as listening to the sounds of the place, experiencing the temperature of the air, registering the color of the light; witnessing the constant movement of people, cars, animals, water, trees and plants; and observing any of the other countless ongoing phenomena of a place.

The 5 second rule serves as a foundation for all of the other methods presented here. At first, like all methods, it can feel a bit awkward or artificial. But after a while, it becomes a seamless part of the process of photography.

Limits

There is one instance when the 5 second rule is not so useful: when the primary subject is constantly moving. For example, a former student of mine studying bees and other pollinators could not effectively use this method while photographing bees feeding on specific plants. But in other situations, using the 5 second rule allowed her, for example, to better see the patterns of plant species in relation to soil moisture.

CONCLUSION

Likely, fewer photographs will be made with this method. Yet memories of and knowledge about the place increase through the intentional act of making time to see and pay attention.[9] As discussed in the introduction, making a lot of photographs does not necessarily yield a good understanding of a place. This method transforms photographing into a process of making, instead of quick mechanical taking. There may be fewer photographs, yet more firsthand experiences will be remembered and synthesized.

My students continue to be surprised and edified by this simple method of making more time:

> I was pretty surprised in the difference I saw in my photos, particularly during the 5 second rule. I feel like my framing and composition improved and I was more thoughtful about making photos than I usually am.
>
> Bella Luikart, student written comments, September 2019

> I liked the 5 second rule because it gave more depth and mindfulness to each individual photograph.
>
> Caroline Schmitz, student written comments, September 2019

> Also "5 Second Rule" reminds me [of] the experience of Film Photographing. I always need to think a lot before clicking the shutter.
>
> Jingyao Wen, student written comments, September 2019

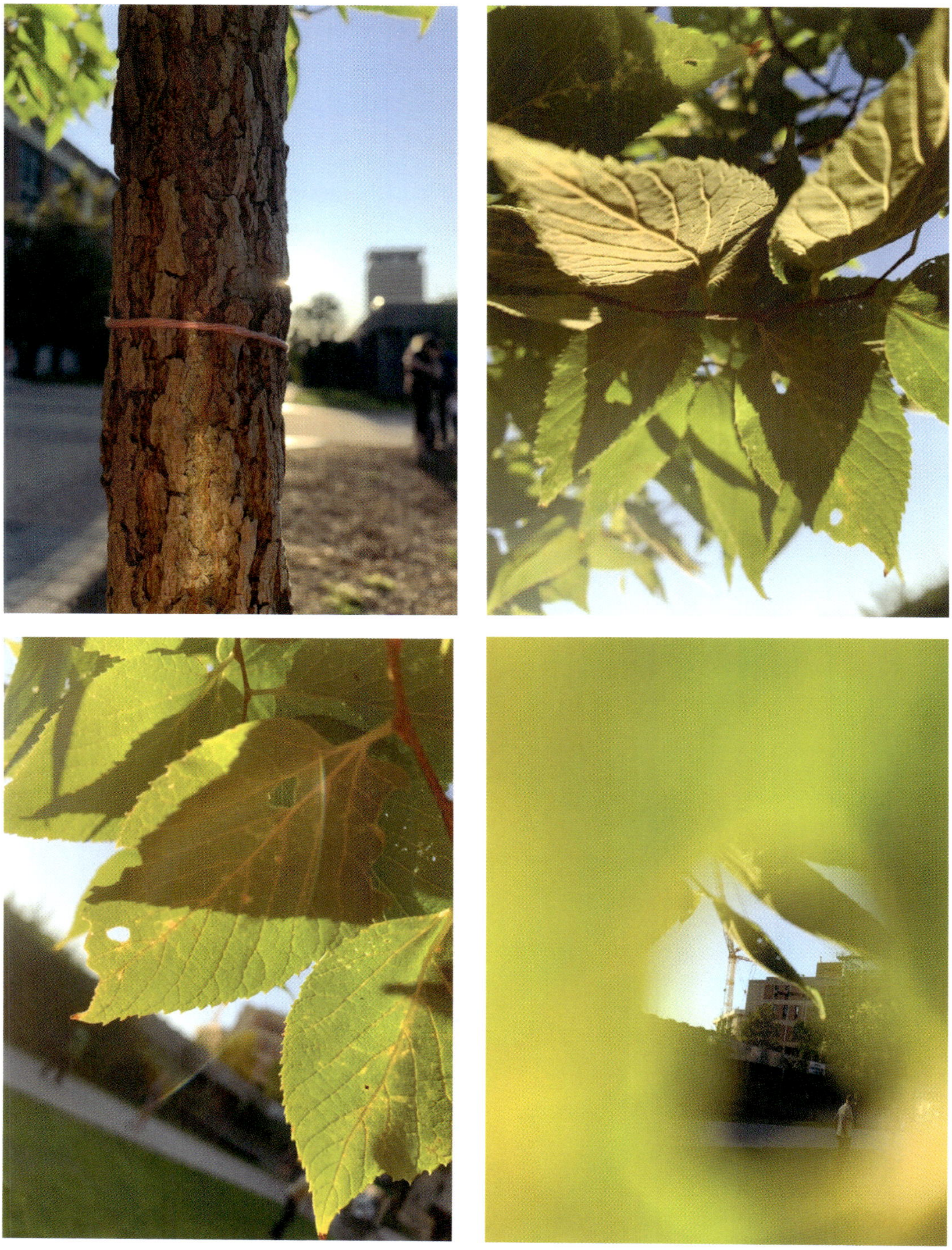

Figures 3–6
The photographer explains her intentions for making these photographs: "I will show the control people have over nature through these photos. The goal is for the viewer to feel nervous at the loss of natural beauty or perhaps amazed at the power of people."

Images: Riley Smith, September 2019.

> My favorite method so far is the 5 second rule, as it really makes you think about what you're photographing. Instead of just taking a quick shot, it's a more thoughtful process, and it avoids taking a handful of pictures and sorting through them; rather I make my adjustments to what I do and don't like before I even take the photo.
>
> Gavin LaFleur, student written comments, November 2019

This method also allows for the photographer to spend more time thinking about what they want the photograph to accomplish. It helps the photographer consciously understand and engage in the condition that photography is a construction that presents ideas to the viewer based on how the photographer composes a scene (see Figures 3–6).

This method forms the basis, both in action and concept, for the set of methods that follow.

NOTES

1 Julia Soares and Benjamin C. Storm, "Forget in a Flash: A Further Investigation of the Photo-Taking-Impairment Effect," *Journal of Applied Research in Memory and Cognition* 7, no. 1 (2018): 154–60, https://doi.org/10.1016/j.jarmac.2017.10.004. This article discusses investigations into what has been coined the photo-taking-impairment-effect. The theory is that taking photographs is a type of cognitive offloading – that the person taking the photograph is less likely to remember what they photographed because the camera is treated as a prosthetic memory tool, instead of relying on one's organic memory.

2 Let's pause to remember that photography used to be something looked at in specialized settings, from sitting and looking at a family album of photographs or viewing stereopairs of distant landscapes as part of after dinner entertainment in the parlor. This kind of photographic looking was done in a temporal and liminal setting. Sharing a photograph was something rare and special. One would give a printed photo to another as a form of affection, remembrance, or act of acknowledgment. Photos cost money to make. The giver of the photograph made a decision to give something away, either their only copy or intentionally having a second copy made (a common practice in the 1990s and early 2000s was ordering "doubles" when sending film in to be printed). For most of the history of photography, the act of giving someone a photograph was an act of giving a gift.

3 Maggie Nelson, "Catherine Opie," in *Aperture Conversations: 1985 to the Present*, eds. Melissa Harris and Michael Famighetti (2017; New York: Aperture, 2018), 415.

4 Evan F. Risko and Sam J. Gilbert, "Cognitive Offloading," *Trends in Cognitive Sciences* 20, no. 9 (August 2016): 676–88.

5 Little Island designed by Heatherwick Studio opened May 2021. Pier55 Inc, "Little Island," last modified May 31, 2021, https://littleisland.org/; Michael Kimmelman, "A New $260 Million Park Floats on the Hudson. It's a Charmer," *The New York Times*, May 20, 2021, www.nytimes.com/2021/05/20/arts/little-island-barry-diller.html.

6 Designed by ARUP and Foster and Partners in collaboration with the artist Anthony Caro. The bridge first opened in 2000 and was reopened in 2002.

7 Anne Godfrey, "Making Photographs: Making Landscapes," in *Active Landscape Photography: Theoretical Groundwork for Landscape Architecture* (London: Routledge, 2020), 33–47.

8 This depends on the type of camera used. A smart phone means looking at a screen view, while using some form of stand-alone camera may mean looking through the viewfinder or rangefinder.

9 This can be referred to as "volitional photography" – when the photographer actively engages in making choices about what to photograph and why they are photographing. Alixandra Barasch et al., "Photographic Memory: The Effects of Volitional Photo-Taking on Memory for Visual and Auditory Aspects of an Experience," *Psychological Science* 28, no. 8 (June 2017): 1056–66.

Method 2 Limits

Helpful with: Getting out of the mindset of treating the camera like a photocopy machine. Making fewer photographs that are more intentional.
Prep: A willingness to stick with the method and see what happens.
Post-production steps: Fewer photographs are made, so more time can be spent looking at each one.

DISCUSSION

It is easy to take hundreds of photographs of any landscape place. Yet do we really need or use all of these photographs? What is lost when we take so many photographs? What can be gained from limiting the number of photographs made?

Quentin Bajac, former chief curator of photography at the Museum of Modern Art discusses the condition of countless photographs and presents one approach to navigating the situation.

> The real problem facing curators and historians of photography is the overflow of images. The main concern for curators in the 1960s, 70s, and even early 80s was to provide access to photographic images. Now we have too much access to too many images. So, we need to be very, very selective. I think it is important now for us as curators to be able to educate the public and direct them in this ocean of images. . . . Sometimes less is really more.[1]

During the research and interviewing process for the first book in this series, *Active Landscape Photography: Theoretical Groundwork for Landscape Architecture*, practitioners voiced a particular approach to site photography over and over:

> *If we just take enough photographs during site visits, we will get what we need.*

This sentiment, in many instances, was quickly followed by an acknowledgment that this might not be the best way to work, but it was the default. These interviews were straightforward and honest discussions about both the benefits and pitfalls of current practices in photography. As the interviews progressed, often rigorous and thoughtful approaches were unearthed and articulated, leading to the realization that specific photographic methods were already embedded in many of the interviewee's site photography practices

DOI: 10.4324/9780429284649-5

for specific design projects. They just needed to be identified and valued as thoughtful methods, and those methods did in fact differ from the "photograph everything" mode. These discussions often included an acknowledgment that "we hadn't really thought about this before."

This led to further exploring the implications of this common approach to site photography. Thinking that if I photograph enough, I will get what I need is frequently joined to the belief that photographs act as a 1 to 1 recording or copy of the visual characteristics of that place (see Figures 1–8).

But this is not true. The act of framing alone alters how humans view a landscape subject in a photograph because the cone of vision itself is reduced and made into a rectangle. This is just one example of how a camera does not act like how we see with our eyes. Any accessible everyday camera lens is a single lens set with a limited aspect ratio. Even multilens cameras on smartphones are programmed to produce a photograph that creates the appearance of a single lens-based image. Our eyes are two constantly moving lenses with both an in-focus cone of vision and an out-of-focus peripheral vision. Photographs do not represent the sensation of peripheral vision. The way we see is not simply a projection of light onto a light-sensitive surface (sensor←camera lens ≠ retinas←eyes). Instead, a very complicated set of neurotransmitters and translations by the brain influence how we see.[2] This complicated concept is a theme examined from multiple points of view in the first book of this series. It is presented again in the introduction to this book and now here.

Even advancements in smartphone cameras, such as the use of multiple lenses and multiple photos taken simultaneously, which are then "chosen" and meshed together based on a set of preprogrammed algorithms to make the "optimal" image, do not result in approximations of how we see with our eyes. The aesthetic preferences that direct the algorithms are articulations of subjective aesthetic ideals programmed by real individual people.[3] These aesthetic choices are not neutral or universal. Thus, the appearance of images made by these programmed cameras is greatly influenced by a set of people making subjective choices about what looks best when a camera digitally processes light capture.[4]

This is all to say that the assumption *If I just make enough photographs, I will get what I need* does not hold up when we realize photographing is not the same as looking and seeing with our own eyes. We can't trade out our eyes for the camera and expect the same thing.

Let's approach this from another angle: what are the results of making "lots" of photographs?

First, those hundreds of photographs need to be sifted through at some later point in time. This sifting is usually fast-clicking or scrolling through these hundreds of photos on some kind of screen. These hundreds, possibly thousands, of photographs quickly become a visual, sensorial, and psychological overload. Rarely is there enough time to thoughtfully and critically look at and see every single one of these hundreds of site photographs. We might even catch ourselves, while out on a site making photographs for a design project, thinking, *I'll look at the photographs more closely later when I have more time.*

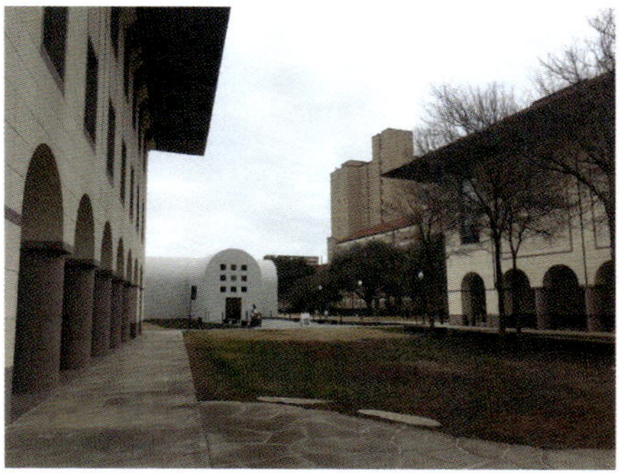
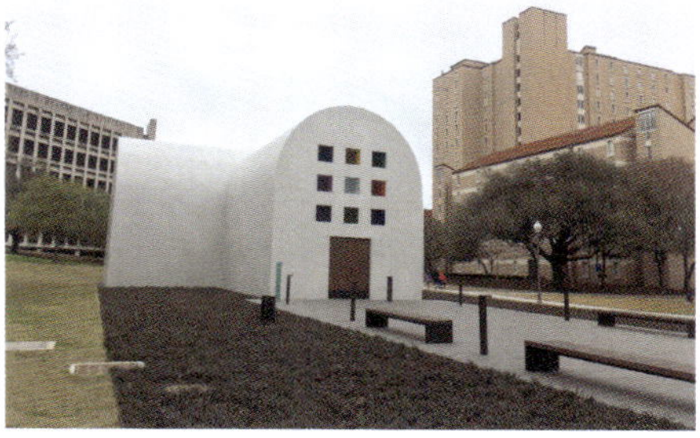
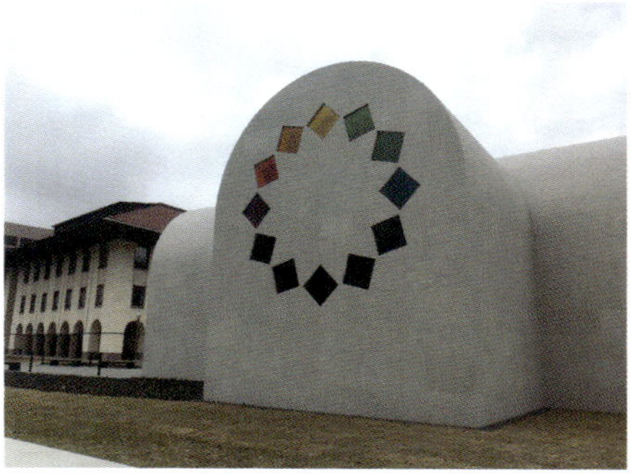
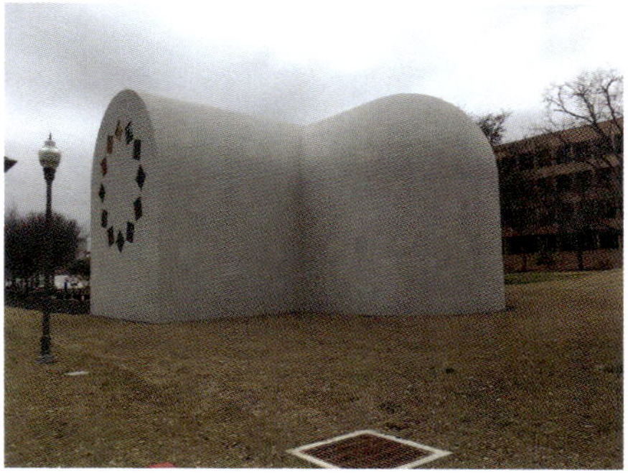
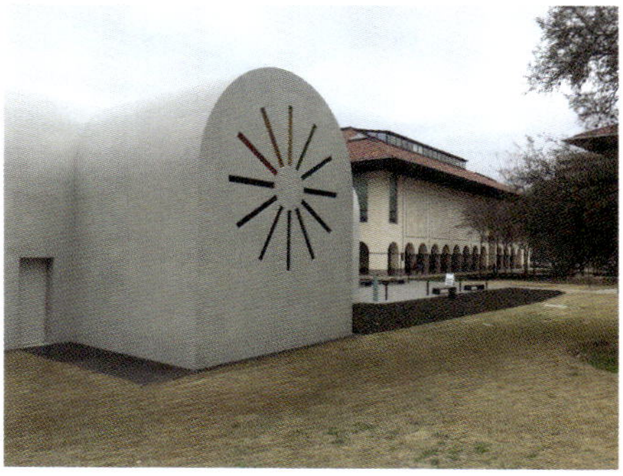
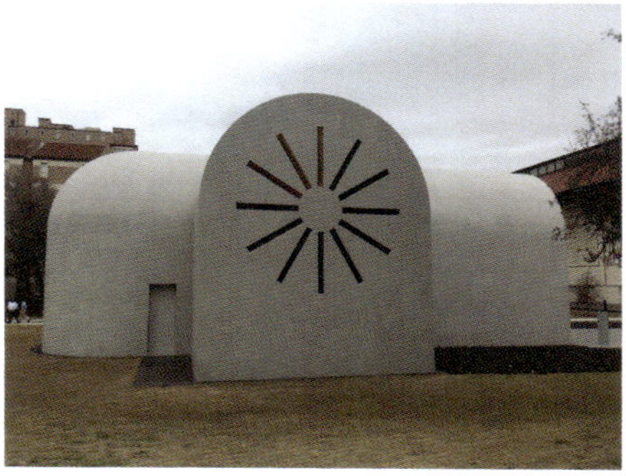
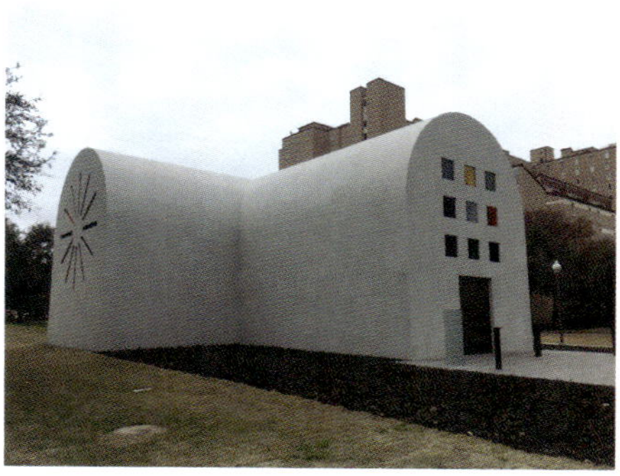
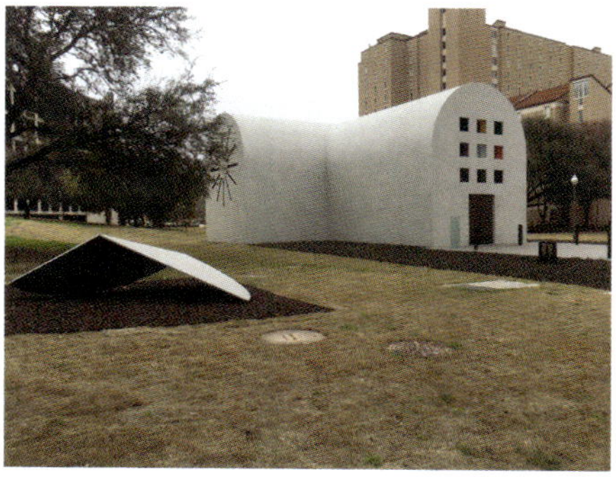

Figures 1–8
In an attempt to assess the placement and context of Ellsworth Kelly's work *Austin* on the University of Texas Austin campus, I made lots of photos. These are just some. In looking back at them, I realize I got into a mode of taking instead of really looking. Enamored of the object form itself, while being confounded by its placement on campus, I treated photography as a shield for looking. In the end, the photographs and my memory of being there are not particularly focused. It appears that I am circling *Austin* but not really landing on anything concrete.

Images: Anne C Godfrey, February 2018.

Second, out of all of those photographs, only a small set become useful for site understanding and design decision making. Each design project collects and relies upon a chosen subset of photographs that are referred to over and over. The rest get tucked away into photo subfiles for the project (on a server or in the cloud). Eventually, all these photos become something to be stored (see Figure 9). This harkens back to the days of dead files where stacks of photographs would be placed in storage, eventually often thrown out.

I notice this tendency to take vast amounts of photographs with my students as well. Observing this phenomenon was one of the primary reasons I started this line of research. I would take the studio group out on a site visit; everyone would tumble out of the van and almost immediately put their cameras in front of their faces and go off photographing.

Diana Balmori refers to this tendency as well. "I have noticed that when students work with a camera, they move about restlessly all the time and are soon ready to move on."[5] Yet Balmori, as do many others, thinks drawing is the antidote to this situation – that the problem lies in the camera itself. But this is not the case. Students are not taught how to photograph or think about photography as a representational medium. And thus, this passive cycle of logic is perpetuated in landscape architecture and our other sister environmental design disciplines:

The camera does all of the work.
The camera creates images that replicate what I see.
Anyone can make a photograph.
No one needs training in how to photograph.
Environmental design disciplines do not regularly teach photography as a complex form of representation.
Photography is blamed for being passive, static or lacking rigor.

This form of thinking scapegoats photography as the reason why the students/designers/photographers don't look and see closely. When really it is our naïve, even lazy, ideas about and use of photography that prevent us from understanding how powerful and complex this representational medium is.[6]

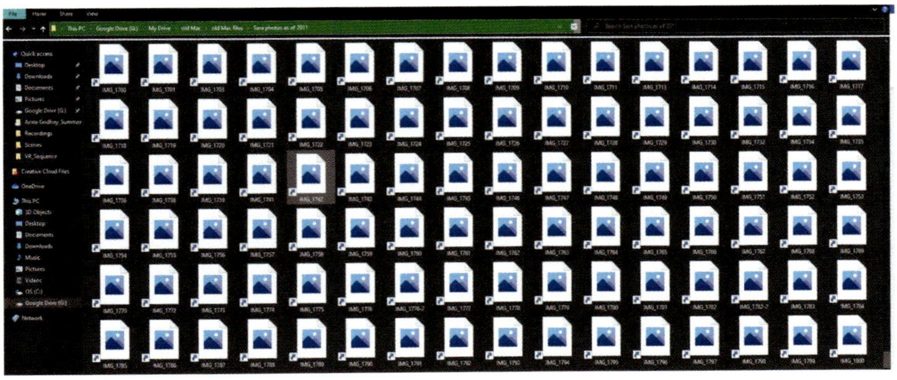

Figure 9
Often, one is faced with sifting through a vast number of site photographs back at the office or in the studio.

Image: Sara Constantineau.

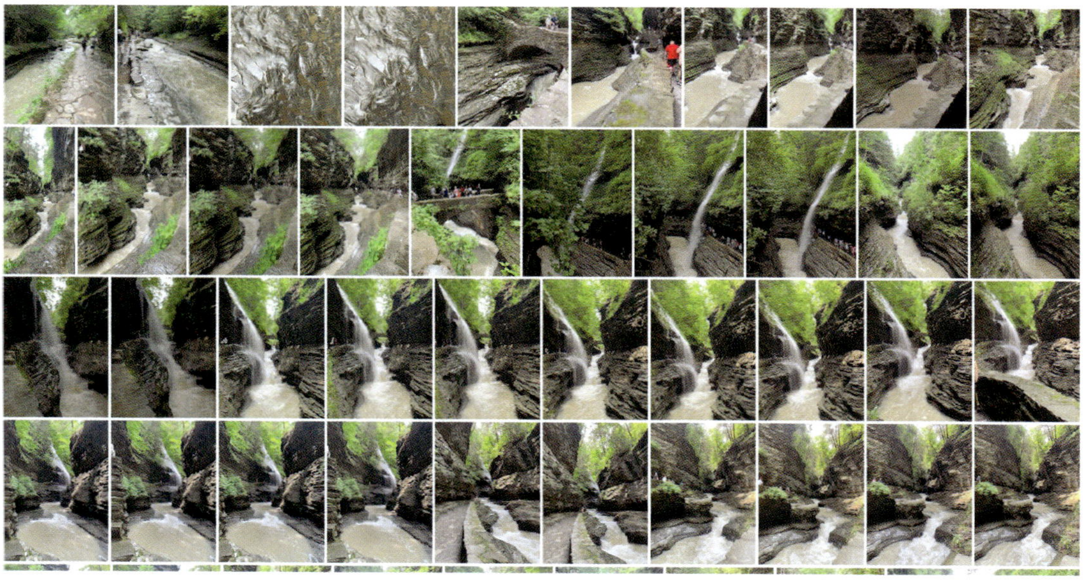

Figure 10
Many (similar) site photographs are taken during project site visits. Here they are viewed in a typical photo library.
Image: Sara Constantineau.

Let's return to the field trip example: once back at the studio, the students would sift through their hundreds of photographs (see Figure 10). This was often done quickly with photos chosen by some gut reaction, not based on a set of pre-stated criteria. I started asking them "how many of these photographs are useful for your design process?" Eventually, the answer became 10–20. Maybe 30 at most for a larger landscape and context.

Observing this trend – making hundreds of photographs but finding only a small subset useful – led me to limit the number of photographs a student could make on a site visit. This common situation became the seed for this research, prompting this question: *How do we better use photography for site investigation and design?* Out of that question came the foundation for many of the methods presented in this book.

THE METHOD

This method is inspired by how a roll of film constrains the number of photographs made. What if I can only make 24 or 36 photographs?[7] What if my budget allowed me to purchase and develop only two rolls of film = approximately 48–72 photographs? (see Figure 11).

This can be done without film, of course. Simply name a limit to the number of photographs one can make during a photographing session. This limit starts to alter the photographic process if it stays under 100 photographs. One's outlook on photographing starts to change. *I have to make choices*. When a small number of frames is allowed, like 24 (back to our roll of film) the process of making critical choices while photographing becomes front and center. Why?

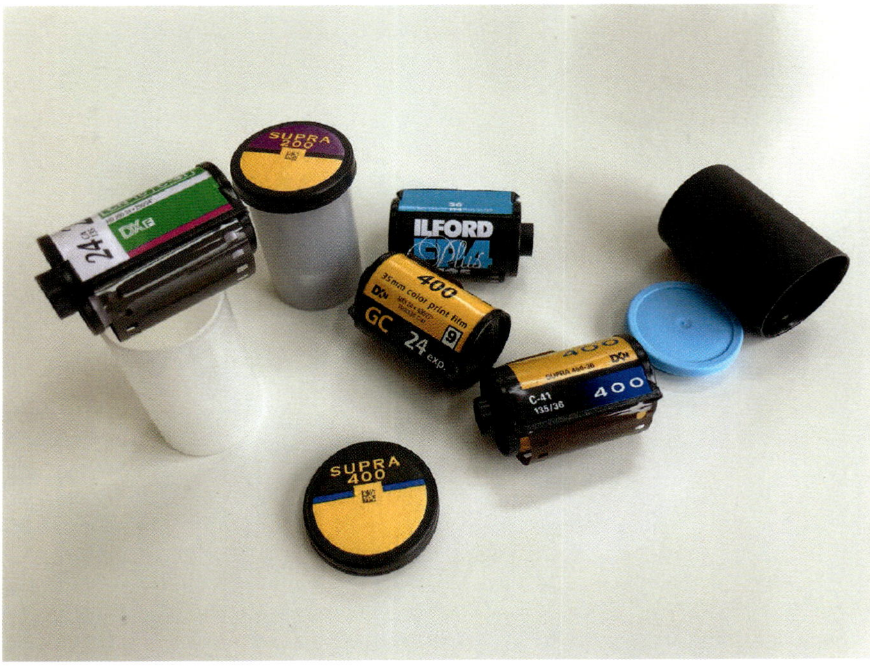

Figure 11
Rolls of 35mm film come in 24 or 36 frames per roll. This provides a good limit for making every photo count.

Image: Anne C Godfrey.

This limit immediately elevates attention. If I can only make 24 photographs, each one of them must help me. Each one has to count. "Make every frame count" is a common prompt in an intro photography course. This limit flips on a switch of critical looking, seeing and thinking. This method introduces the idea that making fewer photographs results in more meaning in each photograph made. One starts to ask, *Why am I making this photograph?* (We will look at this question more closely in the next method.) If every photograph made must mean something and do something, then attentive looking and seeing is activated to support this goal.[8]

Conversely, when we possess the ability to make hundreds of photographs in a short span of time, the camera becomes a shield for seeing. Countless photographs are made thoughtlessly without a real reason for making them. We don't make every photo count. This leads to defaulting to thinking the camera is doing the work. We look less and see less because we treat the camera as a mediator, filter and actual shield for our experience.

Over time I have developed a set of limits in relation to the size of a site:

24 to 36 images for a smaller human-scaled place – like a city block-sized park.
50 photographs for a medium-scaled place – such as a neighborhood.
100 photographs for either a large place or a place that can only be visited once. (This is a type of place that takes 20 minutes to walk across or has multiple subsites.)

It takes some practice to get a feel for this. Try to not give up and default to photographing everything! These limits will change the way you think about photographing. This is challenging – it is meant to be.

CONCLUSION

With a limit, each photographic choice matters. The internal dialogue of assessment is heightened when one uses a limit. It also increases conscious understanding of *Why did I make that photograph?* A stronger relationship between meaning and understanding is cultivated.

Time will feel like it expands when making fewer photographs. Apply the 5 second rule. Look, see, adjust the framing, have a dialogue about why you are making this particular photograph. It will often take more time to photograph when paying attention to the process of making each photograph. For example, making those 100 photographs on a larger site would take several hours or a good portion of a day.

The primary barrier I notice for this method Is the photographer/designer's willingness to really try this out. If you are already concerned that this might not work – *how can I possibly take the risk of using this method out on our next site visit?* – try this: have two people photograph during the visit. One person would be the control: they "photograph as much as possible," while you photograph with a predetermined limit. Compare photographs; see what differences exist between the two sets. Or, practice making a limited number of photographs of a site that is not work related. Think about what happens.

As stated at the beginning of this chapter, this method is inspired by combining the limit of a roll of film with the photography course prompt: *make every photograph count.* Making every photo count does not mean absolutely 100% of the photos made are useful or meaningful photos, but it does lead to learning from what you did.[9] If a photo is not helping you, ask yourself why and consider what you would do next time. It's much easier to have this kind of self-reflection with 36 photographs instead of hundreds of photographs.

NOTES

1 Philip Gefter, "Quentin Bajac," In *Aperture Conversations: 1985 to the Present*, eds. Melissa Harris and Michael Famighetti (2013; New York: Aperture, 2018), 24.
2 See the introduction. Also: The BBC, "Biology KS3 & KS4: How Humans See," accessed June 4, 2021, www.bbc.co.uk/teach/class-clips-video/biology-ks34-how-humans-see/z4wt382.
3 For example, a contemporary discussion: Ian Bogost, "Your Phone Wasn't Built for the Apocalypse," September 11, 2020, www.theatlantic.com/technology/archive/2020/09/camera-phone-wildfire-sky/616279/.
4 This is similar to different chemical and physical qualities found in film or photographic papers. Film and photo papers were and are designed to achieve certain types of aesthetic outcomes. For example, some people prefer Fuji films over Kodak films because of the way blue and green are presented in Fuji's chemical makeup. In terms of choices in photo papers, one particular photo paper may more easily create a higher contrast print, while another paper may display a wider more subtle range of grays. Similarly, color papers "act" differently when printed and processed. For example, I tended to favor AGFA color paper over Kodak color paper, simply because I preferred the depth and color range the chemical makeup of the AGFA paper

produced. (Though to date neither of these companies produce color enlargement cut sheet paper anymore.) These kinds of differences are now found in papers made for digital printing – Hahnemühle produces a wide variety of papers to achieve different purposes, while Moab papers are more focused on landscape photography. Color presentation is different between inkjet color printer brands, and their proprietary inks. (For example, I like Epson printers, but other people prefer Cannon printers.) There are several types of print dialogs utilized to print digital photographs that augment the way printers translate digital images. Some of those dialogs are made for specific papers (ICC profiles), while others attain different types of color outputs. This is just the tip of the iceberg in the realm of digital photography printing. All of this is to say that every step of each of these processes involves making subjective choices about the aesthetics of visual presentation. Yet another set of reasons why photography cannot be equated with seeing with one's eyes.

5 Diana Balmori, *Drawing and Reinventing Landscape* (Chichester: John Wiley & Sons, 2014), 45.

6 I hope that my ongoing research, writing and lectures on this topic will expand this conversation so that at some point I will less frequently encounter one knee-jerk reaction to this work: "Well drawing is better, so we should just do that instead!" This kind of comment is passive and simplistic. And quite frankly, I find this dialog only interested in seeking to cut short diverse discussion instead of critically solving current representational issues within our discipline. Photography is the most important form of representation we use; to keep dismissing it is to continue to remain willfully blind to its pervasive use.

7 The two common number of frames in 35mm roll film.

8 I have instituted these types of limits for my students on hundreds of occasions in countless classes and studios. I see a marked difference in the focus and attention of their photographs. Also, they are often able to describe more fully what they were experiencing on the site while photographing.

9 Richard Misrach refers to this process of self-assessment and learning throughout his book *On Landscape and Meaning*. Richard Misrach, *Richard Misrach on Landscape and Meaning* (New York: Aperture, 2020).

Method 3 Asking Questions

Helpful with: Becoming consciously aware of photographic choices. Examining personal biases that influence site photography. Transcending the tendency to treat photographing as an automatic activity.
Post-production steps: Writing down answers to the questions. Consider having them stay attached to the photographs in some way.
Other materials: A notebook to record thoughts while photographing.

DISCUSSION

When photographing a site or its context as part of the process of analysis and design development, it's easy to react to physical visual relationships without consciously acknowledging the source of those reactions. Photographs are made, yet when looking at them later, the reason for making the photograph in the moment might be lost. *Why did I make that photograph?*

A simple remedy can refocus attention: explicitly asking questions while photographing. This can be done while using the 5 second rule. While looking at the screen or through the viewfinder ask:

Why am I making this photograph?

This can trigger a more conscious and available dialogue about one's interests when working with a site – and more specifically, why one might be framing a particular set of subject matter. The physical visual subjects presented together in the frame of a photograph create relationships (see Figures 1–2). These relationships, as perceived by the photographer, are usually the answer to the question, *Why am I photographing this?* Similar to the previous methods, this method disrupts treating the camera like an automatic self-motivated device for recording that is used in a passive manner. By asking and then answering the question in the moment, more is remembered and the activity of making the photograph becomes conscious and active. I recommend keeping a field notebook and taking a moment to jot down some thoughts after making each photograph. In my classes, this is an explicit assignment for the students.

Working this way, the motivations of the photographer are put front and center, acknowledging that it is *you* who are creating and emphasizing particular relationships

DOI: 10.4324/9780429284649-6

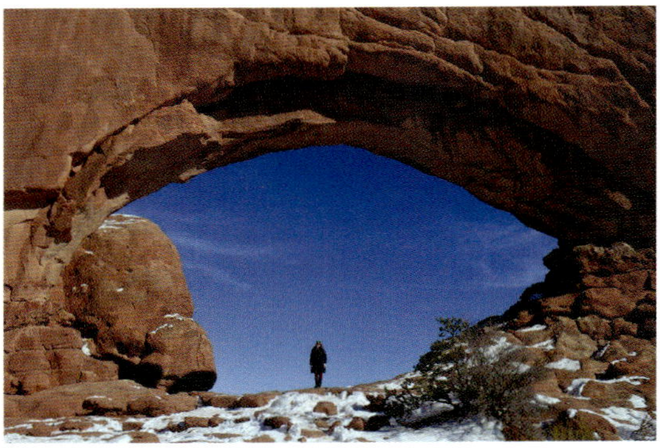
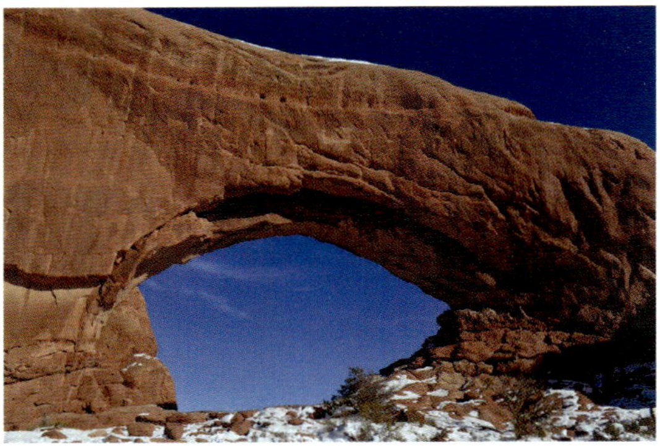

Figures 1–2
The presence or absence of one characteristic can drastically alter what a photograph is about. In Figure 1, the photograph is about the person in this place, Arches National Park outside of Moab, Utah. While in Figure 2, the photograph becomes about a well-known geologic feature of this landscape in relationship to the bright blue sky. One could argue both of these photographs are also about showing the fact that the photographer was there.

Image: Sara Constantineau, 2013.

(and narratives) about the site through how you choose to photograph particular subject matter (see Figures 3–4).

This discussion leads to the second question for this method:

What am I not photographing?

Not photographing is just as important of an activity as photographing. When we do not photograph something, we are stating that it is not of interest or useful. This can be implicit or explicit (to be explored further next). Actively noticing what you are not photographing leads to assessing personal biases. You can assess if you are unintentionally leaving something out and confront yourself as to why that may be the case *while photographing*. Then, the choices for not photographing certain subjects become known and acknowledged.

Asking these questions is not intended to encourage going back to photographing everything (the fallacy of *if I photograph everything I will get what I need*). Instead, it is a way to consciously attach meaning to the photographs that are made while photographing. Additionally, these questions help the photographer identify reasons why other potential subject matter are not photographed.

A more complex augmentation of this method is to articulate an explicit theme or thesis for the making of photographs. This theme could be identified while photographing through the process of answering the two questions, or the theme could be predetermined, based on a stated goal or criteria of the project. The conscious identification of a theme sharpens attention to characteristics that fall within it. Photographing with a theme then acts as a guided form of investigation. For example, I might be looking for traces of past flooding. That photographing session is then consciously acknowledged as focusing on that particular phenomenon.

A third question may be added to this set:

What am I trying to understand?

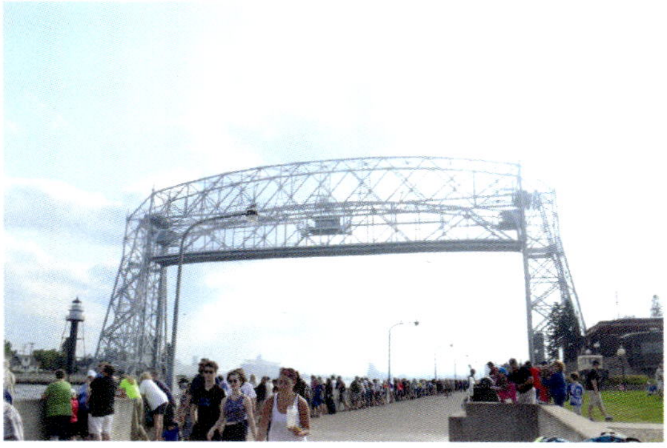

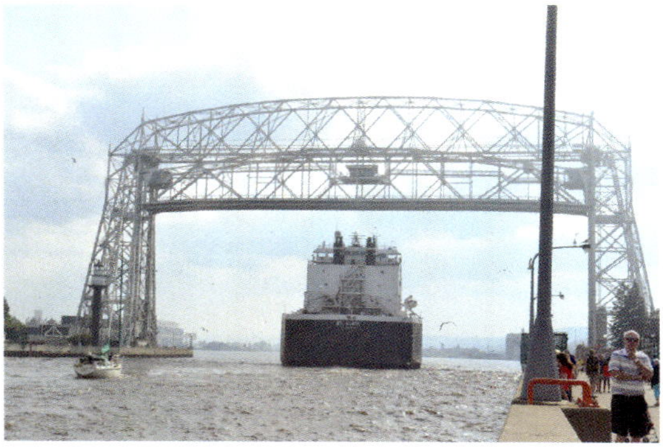

Figures 3–4
Here we see how two photographs in the same place create two different sets of relationships. In Figure 3, the photographer is interested in the patterns of human congregation along the Duluth Lake Walk – people are gathered against the wall to look at the water. While in Figure 4, the photographer presents the spatial relationship between the ship and the Duluth Aerial Lift Bridge (constructed in 1901–1905) that spans the ship canal at Duluth Harbor in Minnesota.

Image: Sara Constantineau, 2014.

Photographer An-My Lê describes her process of asking questions while photographing large landscapes that often present a complex form of conflict:

> I am always asking myself what it is I'm trying to find out about the situation, about the people: "What does it all mean in terms of a cultural and historical context? What are my personal issues here? What is the point? Is it worthwhile? What kind of photograph could attempt to broach and perhaps begin to answer those questions?"[1]

By asking *what am I trying to understand*, a more subtle and nuanced dialogue can be cultivated. Lê's own questions demonstrate how photographing can get at sociocultural aspects related to landscape places.

For example, when I ask myself the first question, *Why am I photographing this?*, my reply might be: *I am showing how the bike lane and bus lane are laid out at this intersection*.

If I ask the third question, *What am I trying to understand?* I might start with more questions: *Is the transportation context for this site something that influences the use of this sculpture in this public space?* Then I make a photograph showing the bus stop at the corner of the site, combined with the bike lane and the crosswalk, and the lanes of traffic headed toward the intersection (see Figures 5–8). I'm trying to understand how the context impacts use. I might make a couple of photographs investigating this issue. Another question may come up, such as, *Is there a conflict with the bus stop and the public space, or is it an attribute?* Then I may make a small set of photographs investigating this idea. Here, we find that snow is not consistently plowed from the designated bus pull-off area. I might take notes about more questions that arise while I photograph: *How often does the bus stop? When the snow isn't plowed, is the bus forced to let people off in the traffic right-of-way? What happens if that is the case?* Then more questions concerning

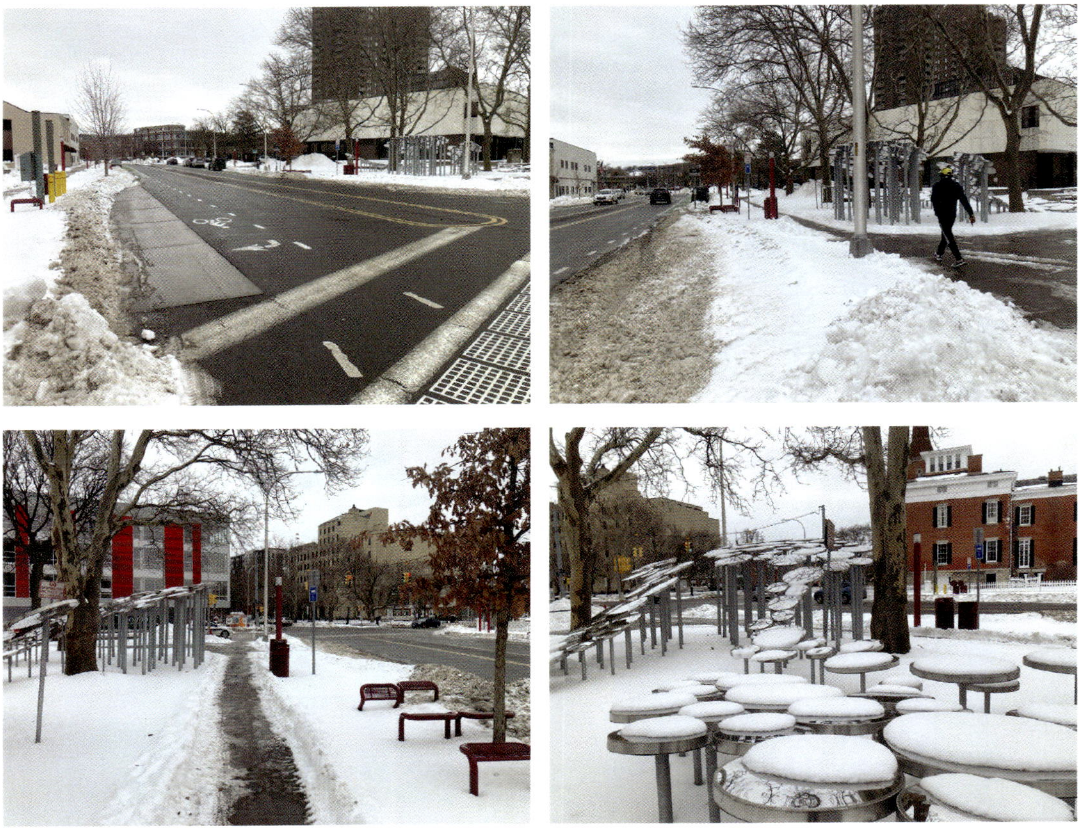

Figures 5–8
The Connective Corridor runs from the SUNY College of Environmental Science and Forestry, Syracuse University and Upstate Medical campuses into downtown Syracuse, New York. This multimodal corridor was designed to establish designated bike lanes and pull-off areas for the bus line. The bus stop here on the corner of South Townsend and East Genesee is adjacent to the public space where "Immersive Cloud" by Daekwon Park sits. Winter time presents an ongoing issue for both public transportation and bikeways – the removal of snow. As can be seen here, the snow has only been partially removed from the designated bus pull-off area, requiring the bus to let riders off in the right-of-way.

Images: Anne C Godfrey, January 2022.

the design arise: *Would the people who take the bus spend time in the public space if there were more seating? If there was a plowed pathway? How might we accommodate these users?*

While photographing, I can also consider how the ways I make these photographs will help other people understand this situation. Through this process of question asking, my choices become more intentional. How do I show important relationships through where I stand, the angle at which I photograph, what I choose to have in the frame, my point of view and what time of day I make this photograph?

Dialoguing with another person while photographing adds more layers to question asking and answering. Together you can discuss what photographs would be helpful to make. Having a second or third person adds more viewpoints to the decision making process. What do they see? What themes do they think about?

CONCLUSION

Combining this method, asking questions, with Method 1 "5 Second Rule" and Method 2 "Limits" results in a useful core set that can be implemented in every photographing session. The 5 second rule creates space for asking and answering questions. Making every photograph count by having a limit to the number of photographs made increases awareness and intentionality. Making each photograph count begs the question "Why am I making this photograph?"

NOTE

1 Hilton Als, "An-My Lê," in *Aperture Conversations: 1985 to the Present*, eds. Melissa Harris and Michael Famighetti (2005; New York: Aperture, 2018), 313.

Method 4 Not Photographing

Helpful with: Thoughtfully looking and seeing before photographing. Creating a stronger and more memorable visual relationship with the subject matter.

Post-production steps: Eventually, this method leads to making photographs. Writing about the experience of not photographing can be part of the process.

Other materials: A way to take notes.

DISCUSSION

As an explicit method, not photographing circumvents automatic thoughtless photograph taking. As referenced earlier, it is easy to just start snapping away without really looking at the characteristics of a site. With this method, the photographer spends time looking at the landscape place they plan to photograph before actually photographing. This looking may be open-ended, without the goal of photography, by simply noticing what is interesting in the landscape.

Or looking without photographing may be a way to test or previsualize photographs. This involves assessing why and what one may want to photograph but not photographing in the moment. Saying *I am not photographing right now* can activate a mode of seeing deeply and bodily. *What might I photograph?* motivates thinking photographically.[1] The photographer makes photographs with her mind as a way to prepare to eventually photograph.

The inspiration for this method comes from two sources. The first is a personal experience of not being allowed to photograph in a place I very much wanted to photograph. The second is the working practices of photographer Jeff Wall, who is discussed briefly here and presented more fully in Part II: Practices.

On a visit to Pennsylvania, I took a tour of home sites by Frank Lloyd Wright, including Fallingwater and Kentuck Knob. The docents explicitly forbade photography of the interiors.[2] Frustrated, I started composing photographs in my mind's eye as we made our way through each of the homes. Photographing with my mind's eye became an active process of investigating. Doing so allowed me to see and better understand some of Wright's design choices. It helped me understand, *in situ*, why I was intrigued with certain aspects of Wright's work. (I was also asking the question, *Why am I photographing this?*) This investigation of photographing with my mind's eye formed a theme about the relationship of volumes of space. I walked around the rooms and put myself in different positions to see the interrelationship of the unfolding spaces. I looked closely at details,

DOI: 10.4324/9780429284649-7

making photographs of those in my mind in a different way – as a kind of note-taking that annotated the larger photographs of space. I would try to be the first person in the room or the last person to leave the room, walking throughout the whole volume to see the space. Though this unnerved the docents at times, my reply was that I was making photographs with my mind. That seemed to help, a bit.

More recently, I was in a similar situation at the Judd Foundation in New York City. The docent-led tour of Donald Judd's five-story cast iron building at 101 Spring Street allowed us to fully experience spaces Judd found generative in the early part of his career (Judd purchased the building in 1968).[3] But we were not allowed to photograph!

On that day, I was interested in how the light gave form to the space, paying attention to the patterns of shadow and sunlight coming through the west and south-facing windows. The light itself created volume. I also photographed with my mind the neat arrangement of cups, glasses and service bowls on the thoughtfully made shelves. I photographed the earthenware bowls under the long worktable. And the Jean Arp drawing in the children's tiny playroom. The docent for this tour was not as alarmed about my wandering through the space as I photographed with my mind.[4] Maybe he was a more laid-back individual or could easily read me and tell I was not going to cause trouble.

Jeff Wall, a conceptually influential photographer practicing in the late 20th and early 21st centuries, is best known for his statement, "I begin by not photographing."[5] Wall decidedly does not participate in reportage – photographing scenes he witnesses in the moment. Instead, when he sees something of interest, he remembers details of it and then makes a photograph later by intricately staging a similar scenario. This remaking of the event is not meant to visually copy what he initially saw; instead, it is a mix of event, compositional and narrative references to painting and literature, and Wall's own photographic intent.[6]

For our purposes, Wall's approach is similar to the method presented here – not photographing – because one looks, sees and identifies characteristics of interest while also considering both its compositional and narrative qualities. A particular instance in a particular place is inspirational to Wall, but he chooses not to photograph in that moment, almost as a way to better understand the situation he witnesses. For our uses, by taking some time to consider what we see and identify what we'd like to photograph, we can better understand our underlying interests in making that particular photograph. In a way, it also disrupts the notion of reportage, as the photographer is scouting the place, noticing, noting and previsualizing particular photographs that will be useful for design decision making – instead of partaking in a more reactive mode of pointing and shooting.

By disrupting the tendency to constantly photograph in the moment, one can more thoughtfully and critically examine subject matter and relationships. We can also start to see and understand compositional tendencies to photograph subject matter in particular ways. Certainly, this method does end in making photographs, but the photographer makes them first without the camera, taking more time to consider the multiple visual, narrative and conceptual factors and characteristics of interest.

When I use this method in my courses, I ask students to walk through a particular site and just look around. No photographing, just looking and being present in the moment. I call them back, and I introduce the idea of photographing with the mind. Then I have them walk through the site again, this time making photographs with their mind's eye.

At first, this method seems awkward and unnecessary. Students sometimes doubt the benefit of this prompt. I get some funny looks. *Isn't this a photography class?* But after engaging the method, they start to see what this is all about. Some express a feeling of freedom when not photographing. Others stated that they saw more. And many said that the photographs they made after not photographing were different than if they had simply started snapping away without first looking.

> It forces you to really observe and then think about the subject matter. And helps to see what aspects of the subject matter you really want to highlight/draw attention to.
>
> Jenna Greco, student written comments, September 2021

> I discovered there is more to look at when we are able to slow down and take a closer look. Also just looking around helped me see things that I wouldn't have seen before. It's peaceful to be able to take in the environment around me and just be in place.
>
> McKenzie Dombroski, student written comments, September 2021

> "Looking without photographing" held the most influence over my experience. This method helped keep me within the bounds of 15 photographs for the assignment. When I was out on the trail, there were many different things that captured my attention. However, after stopping to look through the lens I found the picture did not always hold my interest. This process of reconsidering an image routinely prompted me to implement the other methods into my work.
>
> Raven Bonpietro, student written comments, November 2019

CONCLUSION

By pausing and looking, instead of reacting and photographing, more can be seen and understood. Looking before photographing, asking questions, limiting the number of photographs one makes and taking a bit of time to really look and see before making a photograph all strengthen the relationship between the resulting photographs and the process of investigating and understanding a site.

This method adds a more complex step to the first three methods. Used together, they create a core set to be deployed whenever one photographs.

Looking without photographing
5 second rule
Limiting the number of photographs made
Asking questions

NOTES

1 First published in 1943. Edward Weston, "Seeing Photographically," in *The Complete Photographer* 9, no. 49 (1943). A more accessible appearance is Edward Weston, "Seeing Photographically," in *The Photography Reader*, ed. Liz Wells (London: Routledge, 2002).

2 The following appears under "photography" in the Kentcuk Knob visitor policies. "Exterior photography, painting and sketching are permitted for personal use only. Photography, painting and sketching are not permitted at any point during interior tours." "Prepare for Your Visit," *Kentuck Knob*, accessed June 4, 2021, https://kentuckknob.com/visit/prepare-for-your-visit/.

3 Flavin Judd and Rainer Judd, *Donald Judd Spaces* (Judd Foundation and DelMonico Books/Prestel, 2020).

4 Of historic interest, I made this visit to 101 Spring Street March 11, 2020. I had visited the MoMA JUDD retrospective the day before. The galleries were eerily empty yet allowed me to see Judd's work visually uninterrupted. I made my way over to Spring Street the next day, walked in and was able to join the next tour (unthinkable at any other time). Our small tour group cautiously maintained distance from one another, which was easy to do in the large volumes of the loft floors. By New York standards, the streets were almost empty when I walked out and back to my hotel in the Lower East Side. I met an old friend for dinner. It was the last time I ate inside for over a year. I left the city the next day. Incidentally, that same friend and I ate at the Brooklyn restaurant named Olmsted the following May, which marked the breaking of our year-plus-long hiatus.

5 For example, see *Jeff Wall: "Tattoos and Shadows,"* San Francisco Museum of Modern Art, March 2007, video, 2:21, www.sfmoma.org/watch/jeff-wall-i-begin-by-not-photographing/.

6 See also "Jeff Wall in 'Vancouver'," Art21, accessed December 18, 2020, https://art21.org/watch/art-in-the-twenty-first-century/s8/jeff-wall-in-vancouver-segment/.

Method 5 Quick Simple Things to Pay Attention to While Photographing

Helpful with: Becoming aware of the whole frame while composing a photograph. Varying point of view beyond human-standing-eye-height. Increasing variety in point of view.

Post-production steps: Making and reviewing notes about framing, body position.

Other materials: A notebook or device to keep track of choices made while photographing.

DISCUSSION

The methods in this chapter are an interrelated set of simple and effective ways to alter default modes for framing and point of view. Each is a straightforward prompt often encountered in introductory photography courses.

1) Scanning the Edge of the Frame

Often, the primary subject matter is positioned in the middle of the frame. This is called center weighting. This compositional choice focuses on a single subject while deemphasizing the context. These types of photographs become about the thing-in-the-middle instead of portraying relationships in the landscape.

The tendency to center weight is very common. We point the camera (lens) at something. Like pointing with a finger, the camera is used to single something out. When using the viewfinder (or rangefinder) of a camera, one sights the subject. This creates a singular physical engagement between the camera and subject. I am sighting something and placing it in the middle of my sight.[1]

To counter this tendency, try "running" your eye(s) along the edges of the frame before releasing the shutter. Another way I have come to describe this to students is to imagine tracing the edges of the frame with your eye or your finger. This action is called scanning.

Scanning the edge of the frame heightens awareness of everything in the frame, not just what has been placed in the center. This starts to activate awareness of what contextual relationships exist. The answer to *why am I photographing this?* starts to expand beyond "naming" the object by pointing the camera at it and placing it in the center of the frame.

For example, I might answer the question *why am I photographing this?* with *I am photographing this constructed stream to show that it is here* (see Figure 1). This is an act of naming and pointing. But after I scan the edge of the frame and look at the context

DOI: 10.4324/9780429284649-8

rather than the object, my answer might change. I choose a different point of view, that shows more context. *I am photographing this relationship between the constructed stream and the surrounding plantings while also showing how the features in the stream help aerate the water* (see Figure 2).

Notice the act of scanning the edge of the frame is also an opportunity to reframe the photograph. A stronger visual relationship can be made between the primary subject and secondary context (see Figures 3–4).

2) Move Your Head Around

Rarely do we realize how much we look at and understand our surroundings from a "head looking forward" position. This head looking forward position is usually done while standing.[2] I call this common point of view "human-standing-eye-height." Think about how many photographs are made from this point of view.

But what if we moved our heads around?

Before making a photograph, take a moment to move your head – look up, down and side to side. This allows you to see more of what is around the primary subject (this subject may be an object, a set of landscape relationships in a place or even a long wide view). Taking a moment to look around quickly creates an awareness of the larger context. What is here, what is happening? This is another example of how photography can be treated as a process that both produces photographs and increases experience of a place. If I take a moment to simply look around, even if I do not photograph all of those views, I will increase my experience of the site.

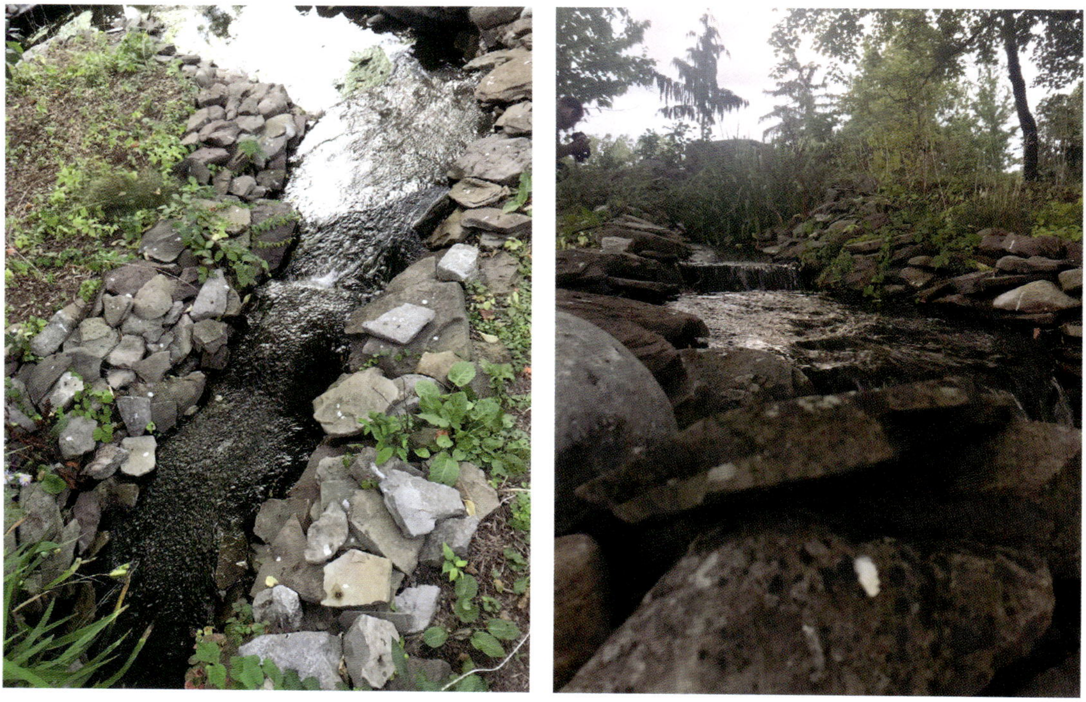

Figures 1–2
Constructed stream and relationship with the surrounding plantings.

Images: Alexander Roukis, September 2019.

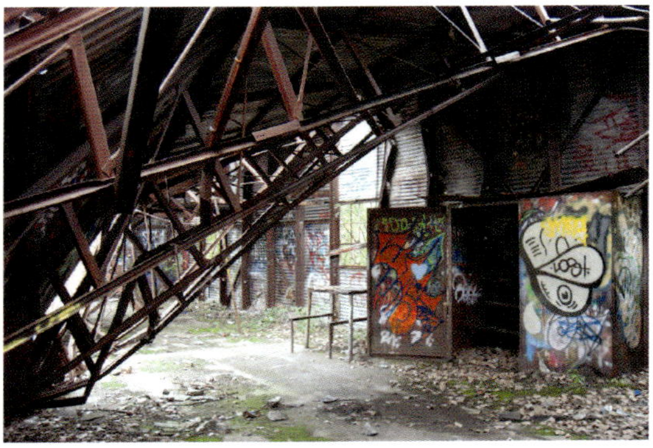

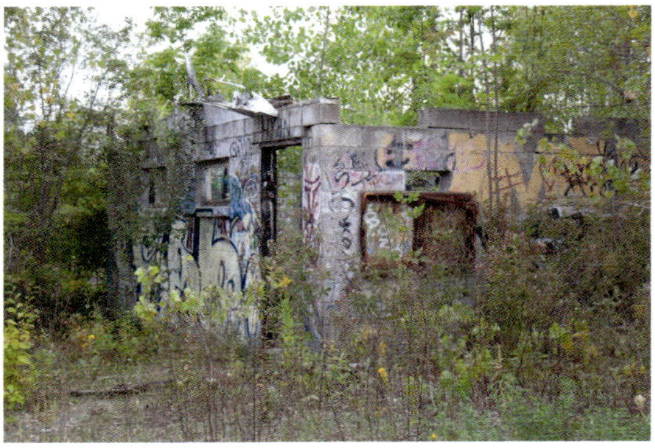

Figures 3–4
Graduate student Autumn Elniski comments on her process of photographing this abandoned, yet often visited, industrial site outside of Syracuse, New York:

"My favorite method thus far is scanning. I find myself trying to use it more and more in my everyday photography on my smartphone. I think it is really useful because you can take a few seconds to see if there is anything in the edges of your frame that could take away from or add to what you're trying to capture. It wasn't something I thought of often, with our usual focus on the center of the frame. In one of the shots . . . I actually scanned the edges and saw that I could include some of the plants growing around the building if I adjusted the zoom just a tad. It really added to the [photo]."

Images: Autumn Elniski, September 2019.

In many ways, what photographer Sally Mann does with her 8 x 10 view camera is similar to moving the head around. An 8 x 10 view camera often requires the use of a "hood" to view the scene projected onto the glass at the back of the camera. Mann describes her process: "Under the dark cloth I rotate the camera on the tripod and watch the randomly edited scenes pass across the milky rectangle."[3] She does this to see and test different possibilities for photographs, instead of quickly settling on a single scene. This kind of work with the camera helps to hone sight and prepare the photographer to be ready for different opportunities while photographing, instead of just sticking to that first tendency – in our case to center weight one particular subject from a human-standing-eye-height point of view.

Stephen Shore discusses this kind of self-training to increase preparedness as well. "It's this funny or odd play between being responsive to what one encounters but having a certain mental framework that digests it."[4] In Shore's case, he "trained" himself to be open to opportunities in vernacular landscapes while applying his particular style of composition to making the final photographs. This approach is similar in many ways to the practices of landscape architecture, where we want to be open to the uniqueness of a place but have developed a tool kit for how to approach certain situations or typologies.

3) Move Your Body Around

The next step is adding your body to this practice of moving around. Try crouching or climbing up on something. Consider making a comparative photographic set with three different body positions: human-standing-eye-height, then one crouching and then one from an elevated position, such as from the back of a truck's tail gate or a park bench (see Figures 5–7). Notice what changes and what remains the same. Think about how moving body position alters the understanding of the place. Scale is one good example

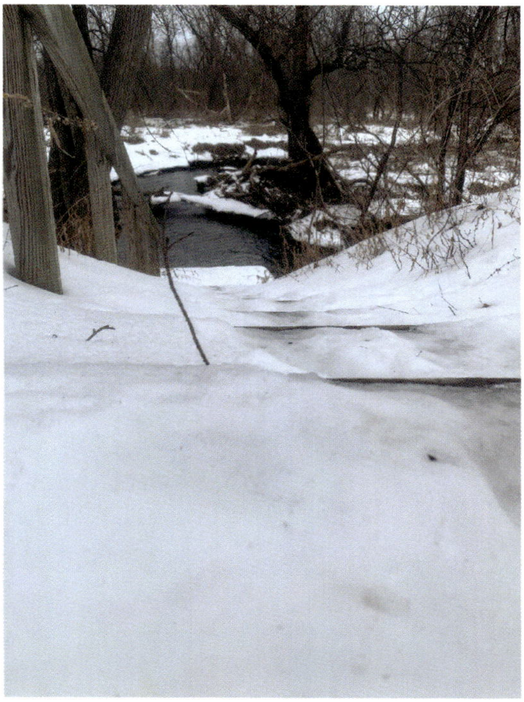

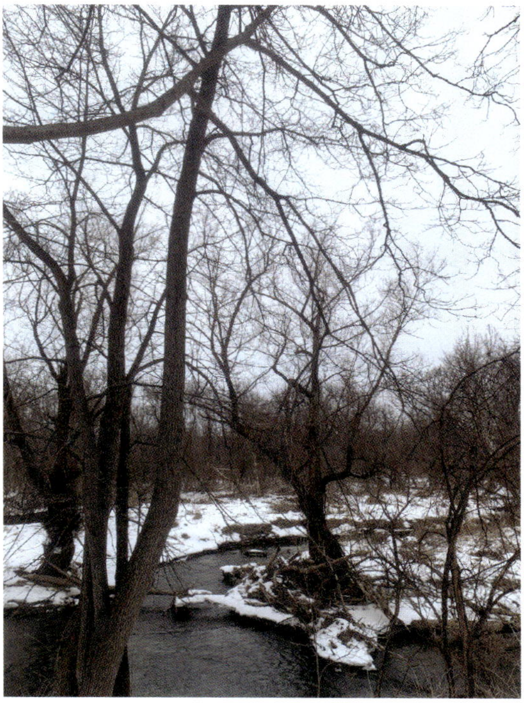

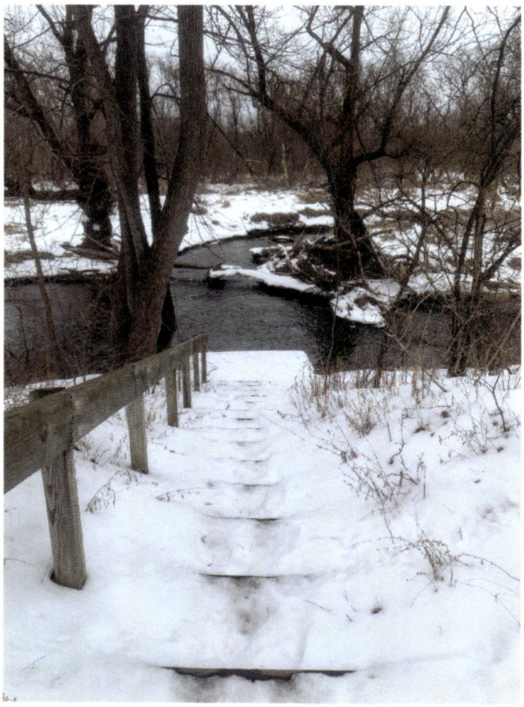

Figures 5–7
Nine Mile Creek, Amboy, New York.

Images: Sara Constantineau, 2014.

of this. The scaler relationships between subject matter can change quite a bit based on different body positions, or points of view.

These physical activities are a way to engage the relationship between what you see with your eyes, what you experience with your body, what you are thinking about on the site and, finally, how you use the camera to interpret these experiences.

A more complex point of view method appears later in Part I.

Limits

For some, these physical movements may not be easily achievable. I may have a different range of movement because I am differently abled. One way to engage in these types of methods is to move the camera around with some kind of extended tool, like a handheld tripod, a monopod, selfie stick, a gimbal or some other device. This might limit the ability to see out of the viewfinder (unless you have a camera that has an articulated viewfinder or use a secondary viewing screen), yet the process of trying other positions with the camera still sets the imagination going in a new direction. *What will my subject look like if I photograph it from an elevated position? Did the photograph turn out the way I thought? What did I learn from this?* Upon reflection, perhaps this is not a limit but simply the development of another form of this method.

CONCLUSION

In combination, these three methods are simple ways to expand the options for framing and points of view beyond center-weighted, forward-looking human-standing-eye-height. The following methods present more specific and complex augmentations to these basic prompts.

NOTES

1 Yes, this does end up being another hunting metaphor for photography. I sight something and place it in the center of my crosshairs. In some camera devices, one can enable different framing marks to show up in the viewfinder or screen to assist in composition. One option in some devices is an actual sight or set of concentric circles that create a bull's eye.

2 I'll take a moment here to point out that standing is still the default position for designing. This is to acknowledge that differently abled people's experience of space (for example, those who may not usually stand) are not often part of the initial way we think about designing a space.

3 Melissa Harris, "Sally Mann," in *Aperture Conversations: 1985 to the Present*, eds. Melissa Harris and Michael Famighetti (1995; New York: Aperture, 2018), 326.

4 Lynne Tillman, "Stephen Shore," in *Aperture Conversations: 1985 to the Present*, eds. Melissa Harris and Michael Famighetti (2004; New York: Aperture, 2018), 515.

Method 6 Frame: Reframe

Helpful with: Practicing multiple ways to frame a subject. Becoming aware of how visual relationships presented through the act of framing influence understanding of the primary subject matter.
Post-production steps: Side-by-side viewing of the initial framing and the reframing.
Other materials: Reframing can be done with the camera or in post-production by cropping with software. Or reframing in post-production can be done manually with printouts and the use of paper masks, tape or scissors.

DISCUSSION

The frame defines what we see – how one frames determines the information conveyed and the sense of space within the photograph. The frame controls the amount of information the viewer is given and how the viewer focuses on that information. Framing can re-enforce or negate a sense of more "things going on" outside of the frame (open framing vs. closed framing). Framing is an act of choosing what is and is not of interest.[1]

To more completely understand the power of framing, this method asks the photographer to frame and then reframe the subject matter of interest in multiple ways. The photographer experiments as a way to understand the power of framing. By making and then viewing alternative framings, the photographer is able to see how perception of the subject matter can be shifted through framing.

As landscape architects and environmental designers, it is essential to think about both what is in the frame and what is outside of the frame. We work with relationships. Usually, framing of a landscape subject cannot show all parts of a landscape relationship – be it cultural, historical, or ecological. For this reason, framing can become an act of abstraction, to a greater or lesser degree, or an act of highlighting and weighting.

THE METHOD

There are several iterations for this method. Three are outlined here.

1) Reframing While Photographing

While out photographing (and utilizing the core methods: 5 second rule, limiting the number of photographs made, asking yourself questions and looking without photographing) make two differently framed photographs of the same subject matter.

DOI: 10.4324/9780429284649-9

Start with your initial inclinations for framing. Make that photograph. Then, look at the landscape and consider another possibility. If you are able to, also look at the first photograph you made to consider different ways of reframing it. Move your head around, or move your body into a position other than human-standing-eye-height. Engage in more question asking. Reframe, then make another photograph.

Make two photographs for every photograph you make (be it on a site visit, or as part of a practice session or while visiting a place you enjoy). Still consider limits. For example, if you are using a 24-photo limit, you would then double it to 48. If you have more time, or more inclination, you could reframe a third time as well.

2) Applying a Predetermined Reframe Rule

Here, the method is to pre-choose, or predetermine the reframing, and then use that predetermined rule throughout the whole photographic session. As mentioned previously, make the first photograph based on your initial inclinations (while using the core methods); then apply one of the following rules for the reframing.

a) The reframe shows 50% less space (zooming in or getting closer).
b) The reframe shows 50% more space (zooming out or backing up).
c) The reframe is taken from an entirely different point of view.

These are just three examples of countless predetermined options (see Figures 1–4).

3) Reframe in Post-Production

For this approach, the reframing is done through the act of cropping, either in a digital format or physically with paper masks, markers, tape or scissors. This method is different because only the available information in the photograph can be used. Yet it is still an effective way to understand how space and relationships can be radically changed by simply cropping out a particular physical/visual phenomenon.

This reframing can be done by the person who made the photographs, yet it is also quite interesting to have the reframing done by someone else. I have presented this activity countless times in courses and studios, asking students to get into small groups and reframe one another's photographs. Taking time for conversations, where each person shares what they were trying to accomplish through the act of reframing, is another opportunity to learn how people see and focus on different kinds of subject matter and see different kinds of relationships in a single photograph.

CONCLUSION

Framing is the most powerful compositional choice in the photography toolbox. Framing designates importance. Framing is the first act of composition and framing creates the boundary for the subject matter presented.

This reframing process is useful in several settings. In the classroom, it becomes a way to understand how people see or value certain things in a landscape. In the office, it can be a way to understand how a colleague's expertise motivates her to focus on particular subjects, while another colleague, with another area of expertise, might focus

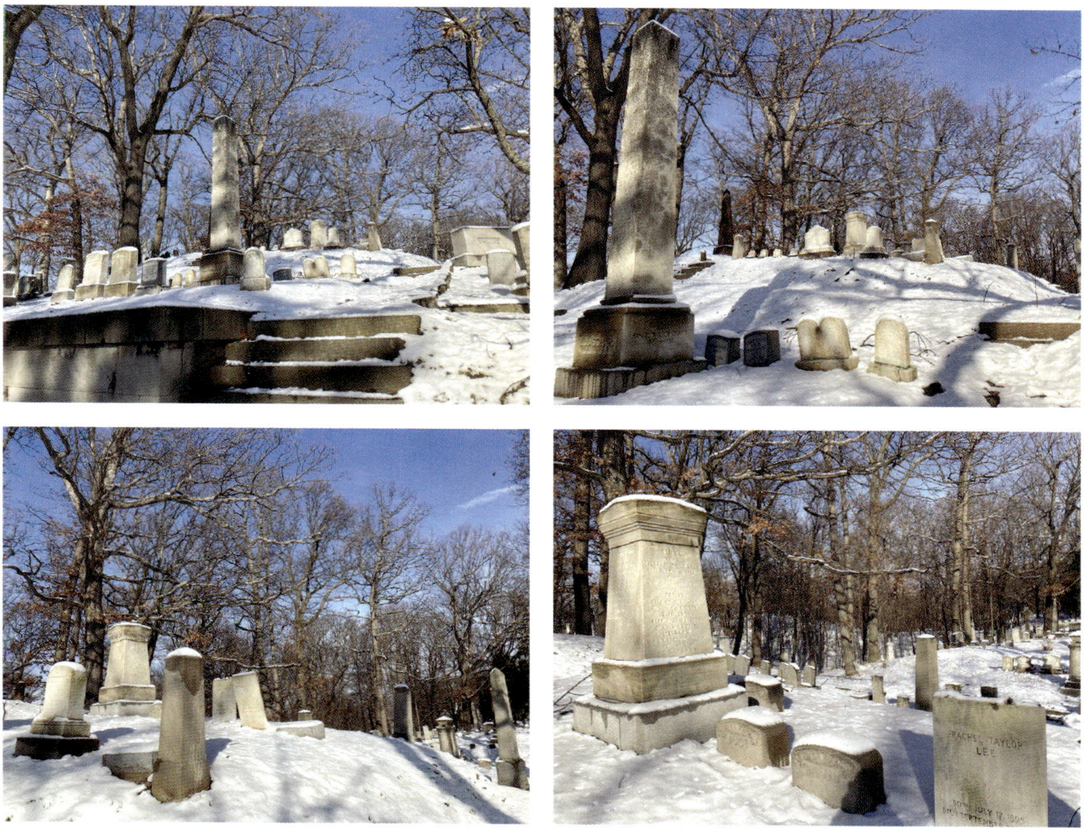

Figures 1–4
In this set, each of the framing choices includes a squat tapered pillar tombstone, which appears in the center of both Figures 1 and 2. For this example of reframing, I physically move my body position to change the subject matter that appears in each frame while choosing to keep the tapered pillar as a focal point in all of the photos. Each image gives a different sense of space and reveals different characteristics of the context. Though the primary focal point is the same, it moves into the left part of the frame in Figures 3 and 4. Moreover, each image is made to present different aspects of the context. For example, Figure 4 presents context that can be seen standing on the rise of this hill, which is not discernable in Figure 1. Photographed at Mount Hope Cemetery, Rochester, New York.

Images: Anne C Godfrey, January 2022.

on different relationships or subjects. When working with communities, clients or stakeholders, this can become an activity for sharing stories and ideas about characteristics and relationships in a landscape.

NOTE

1 Framing is one of the most referred to concepts in photography literature. Two seminal works that pointedly discuss frame are the following:
John Szarkowski, *The Photographer's Eye* (New York: Museum of Modern Art, 1966).
Stephen Shore, *The Nature of Photographs* (Baltimore: The Johns Hopkins University Press, 1998).

Method 7 Compositional Sets

Horizon Line, Focal Point, Near and Far

Helpful with: Understanding how the placement of the horizon line organizes composition. Becoming aware of how the focal point influences visual relationships within the frame of the photograph.

Post-production steps: Displaying the sets together to compare and contrast the resulting photographs.

Other materials: One could diagram the photographs to further analyze the relationships of the focal point and horizon line within the photograph. This could be done by hand with trace or with any kind of software.

DISCUSSION AND METHODS

These methods can be treated as options for site exploration, or they can be used independently as a training exercise to become more attentive to the visual ramifications of the horizon line and focal point.

Horizon Line

Because our subject matter is the landscape, the horizon line usually appears in our photographs. The horizon line can be described as the "line" signaling where the sky and ground "meet." As humans, we read that "line" and are reliant upon it for orientation. The horizon line is often a primary organizing element in a photograph (see Figures 1–2).

The horizon line is visually discernable because the sky and ground appear in high contrast to one another, creating an intersection or abutment. Although in certain lighting conditions, the visual abutment of sky and ground are not so highly contrasted, instead producing a more subtle delineation. Yet, humans still have a sense of it, as the horizon line is always in relationship to eye-height when one is physically navigating the landscape.

How the camera is pointed determines where the horizon line is located in the frame of the photograph. Making photographs from human-standing-eye-height (or within its zone, slightly above or below) often results in the horizon line being placed in the middle third of the picture plane, which is the default position.

What if we were to experiment with pointing the camera at different angles, or point of view, to place the horizon line in other parts of the frame? Issues of scale, the amount of visual space certain features take up in the frame and the understanding of context are all affected by where the horizon line is placed in the image. Placement of the horizon line also determines what is focused on, and thus guides the understanding of different landscape relationships.

DOI: 10.4324/9780429284649-10

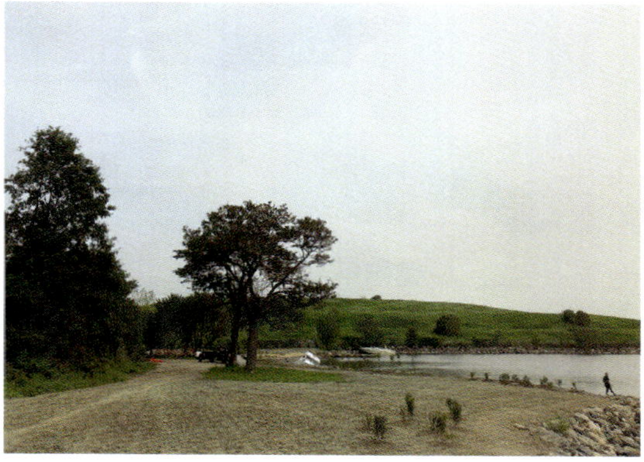

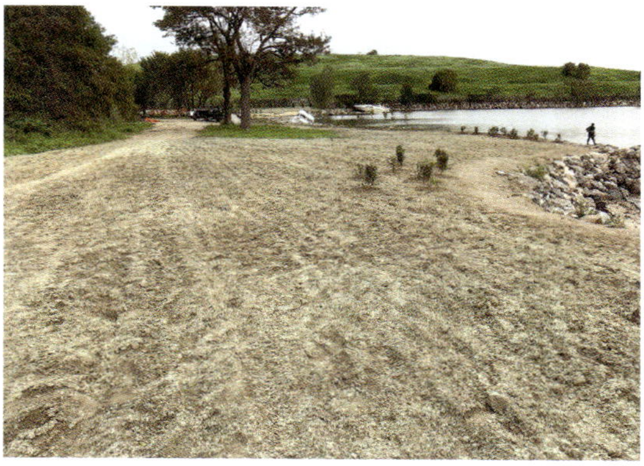

Figures 1–2
The first image places the horizon line in the default position, human-standing-eye-height. For this placement, the foreground and midground take up a similar amount of visual space, while a larger amount of sky makes up the background. The silhouette of the tree and hillside are in high contrast to the sky, and their forms are complete and legible. The second image places the horizon line much higher in the frame, which emphasizes the texture of the ground plane and changes how the space appears. The eye stays toward the ground and becomes more interested in the shoreline in the midground.

Images: Milagros Castelan, September 2020.

By intentionally changing the placement of the horizon line and then comparing the photographs to one another, we can train ourselves to be more sensitive to the implications of this choice. This leads to the usefulness of exploring multiple options for the placement of the horizon line and then examining their effect on the appearance and understanding of the subject matter. A set of 3 or 4 photographs will be made for each landscape scene photographed.

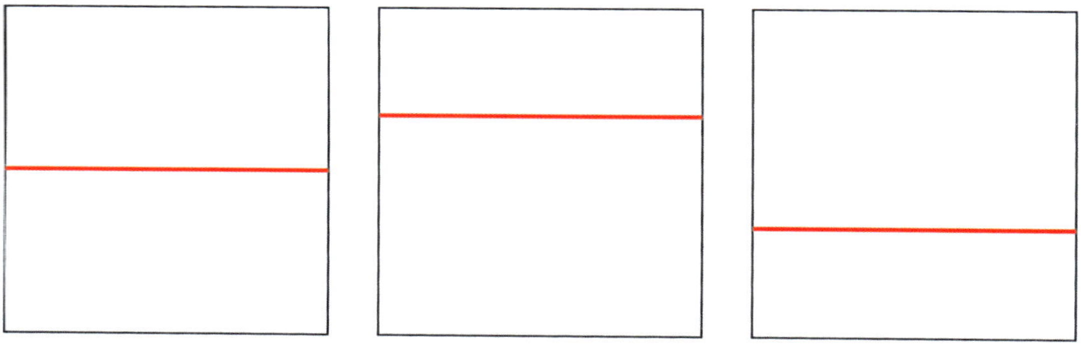

Figure 3
The horizon line can be intentionally positioned in the middle, upper or lower part of the frame to create different visual effects.

Horizon Line Methods

Rule of Thirds

For this method, the rule of thirds is used to organize the frame. The first photograph made places the horizon line in the middle of the frame. The second photograph places the horizon line in the lower portion of the frame, and the third photograph places it in the upper portion of the frame (see Figure 3).

This can be achieved in two ways: (1) tilting the camera down and up to place the horizon line and (2) moving the whole body in ways that place the horizon line lower (crouching, kneeling) or higher (standing on something or using some form of tool to elevate the camera). Though it may seem obvious, it is important to note each of these placements will produce different visual results.

Rule of Fourths

The same process as noted earlier is applied, but instead, the frame is divided into four parts. Each photo places the horizon line in one of the horizontal quarters of the frame.

Not all photographs are interested in horizon line as a primary organizational element. Then what organizes the frame?

Focal Point

Another way photographs are visually organized is through the use of focal point. The focal point is where your eyes go first when looking at a photograph. Usually, the eyes return over and over to that focal point. Focal point can be created by multiple visual conditions: a subject that is most visually stimulating, a subject that is the largest in scale, a subject that is made of angles that attract the eyes to look at it, a subject that is in high contrast to its surroundings (this can be value or color) or a subject that is conventionally different from the rest of the parts of the landscape (one-of-these-things-is-not-like-the-other; see Figures 4–6).

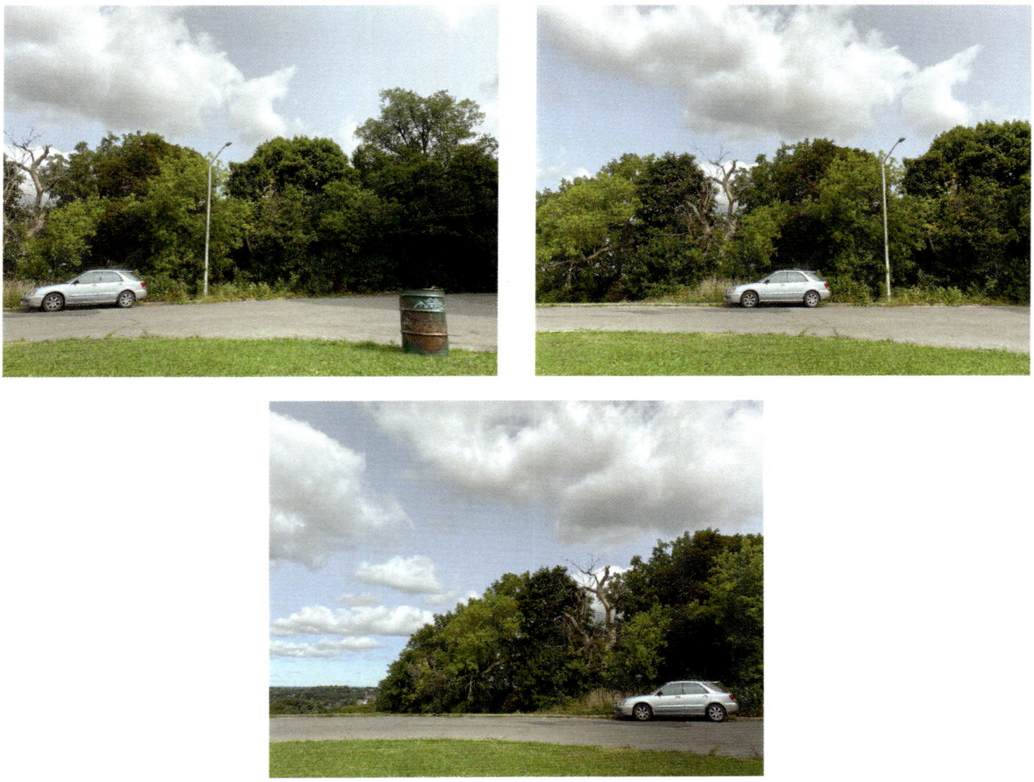

Figures 4–6
Changing the placement of the focal point, this car that stands in contrast to the trees, alters the sence of space and context. The first image presents a closed space, with little sense of the larger context. As the focal point is placed in a different part of the frame, one discovers that this site is considerably higher in elevation than its surroundings.

Image: Devin Reiss, September 2020.

The default placement of the focal point is often the center of the frame. This is called "center weighting" (see Method 5). The placement of the focal point in the frame also influences how we see and understand the contextual relationships within the photograph.

This gets back to the point that photographic meaning is created by an unbreakable four-part relationship between the photographer, the subject, the viewer and the context in which the photograph is viewed (see the introduction). In this case, the photographer is choosing where the focal point appears in the frame of the photograph. This influences how the viewer sees, for example, the context of the primary subject and, of course, what is in the photograph as a whole. The viewer creates narratives about the photograph, which are in part influenced by the composition of the photograph, including placement of the focal point. This viewing occurs in a particular context, for our purposes, within the context of making choices during design development. The placement of the horizon line similarly plays into the unfolding of this four-part relationship of photographic meaning.

Focal Point Method

The focal point will be placed in different parts of the frame to see how this changes the way the subject matter appears and is then interpreted. The same physical movements to place the horizon line will be used here – either tilting the camera or moving the body to change the placement of the focal point in the frame.

This method utilizes the rule of thirds to divide the plane of the photograph into nine parts (see Figure 7). For this method, nine photographs are made, placing the focal point in each of the nine zones of the photograph. (The rule of fourths could be used as well, but that often makes the focal point quite small. Alternatively, one could use the rule of halves.)

There are many variations within this. The focal point could only appear within the space of the individual zone. Or, the focal point could be larger in scale, bleeding into the other adjacent zones, yet still be clearly visually weighted in its primary zone. Or there may be a focal point within your object's subject matter (like an eye or nose). Then the object's subject matter fills most of the frame, and the focal point (eye or nose) position is placed in each of the nine zones of the frame.

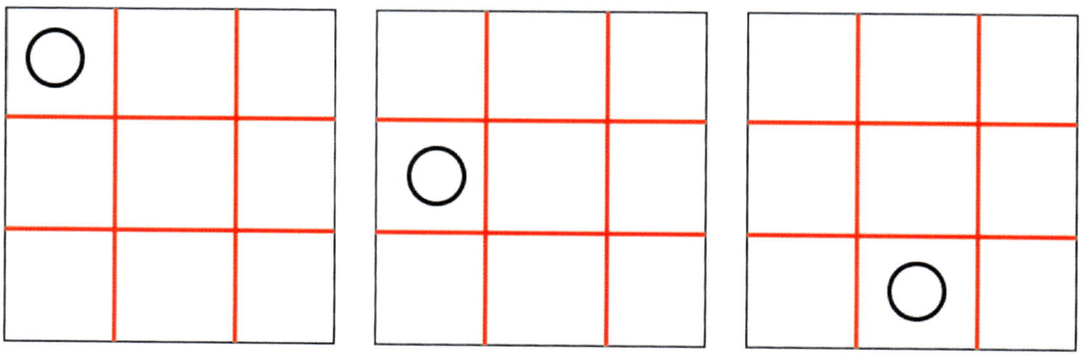

Figure 7
The focal point will be placed in different parts of the frame to see how this changes the way the subject matter appears and is then interpreted. This method utilizes the rule of thirds to divide the plane of the photograph into nine parts. For this method, nine photographs are made, placing the focal point in each of the nine zones of the photograph.

Near and Far

Finally, the physical proximity of the photographer to the primary subject of interest in a photograph determines the understanding of context and the relationship of that primary subject to its surroundings. Often, a subject is photographed from one distance, and then the photographer moves on to the next situation of interest. What happens when we systematically make several photographs of the same subject from multiple distances?

Near and Far was developed in response to watching students photograph the primary subject in what I call the "middle landscape" point of view. Not only is the primary subject center-weighted, but the photographer also stands at a distance from the subject that is not too near, nor too far away, thus the middle landscape view. The middle view often represents human-scaled interests but does not represent larger contexts or more specific details. As a way to get students to see the implications of all of these middle landscape views, I ask them to make 3 to 5 photographs at different distances from their primary subject matter (Figures 8–11).

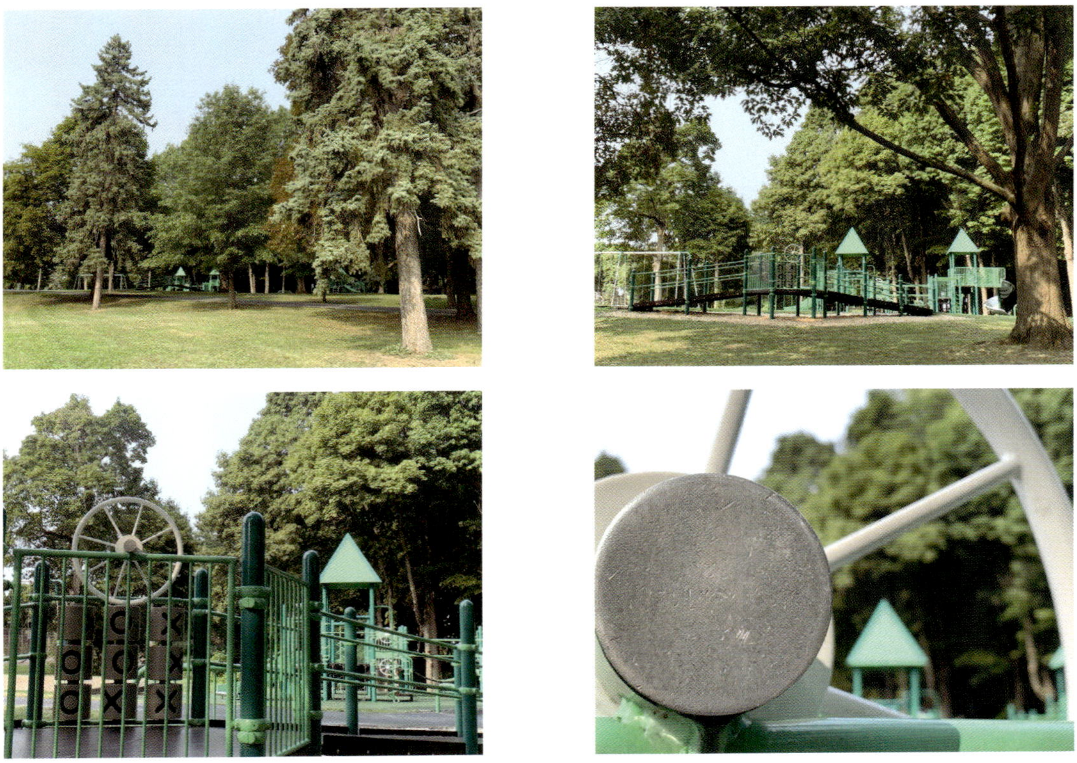

Figures 8–11
By slowly moving closer to the primary subject, the relationships shift and change to the point that the first photograph of this set is concerned with a series of relationships that do not visually exist in the last photograph. In many ways, the last photograph is about something quite different to the first photograph. This method can become a playful way to experiment with the representation of scale and context.

Images: 8–11: Nathan Kettler, September 2020.

In comparison, we see that relationships change when the distance changes. Some characteristics become insignificant, while others become known and apparent as the photographer moves closer or further away. As discussed below, this requires physically moving, not zooming the lens in or out.

Near and Far Method

Once the photographer has identified a photograph/subject of interest and made the initial photograph, a set of photographs are made both closer and further away from the first photograph.

This is different than zooming in and out with the lens. Here, the goal is that the photographer bodily moves to make the new photographs. This becomes another way to navigate and increase the experience of the landscape (see Practice 2 "Predetermined Path").

The photographer may either use predetermined intervals – such as feet, meters or steps – or move closer or further away from the subject at will. At least three photographs are made, the primary photograph, then one closer and then one further away. Often, I find students wanting to make more than three, as they see how much starts to change between each interval.

CONCLUSION

For all of these methods – horizon line, focal point, near and far – the process of making the photographs and the process of examining the sets afterward are opportunities to understand how these compositional choices influence the visual presentation of the primary subject matter and context.

These methods demonstrate a wider palette of compositional possibilities, helping the photographer/designer transcend tendencies to only photograph in particular ways, with particular compositional defaults. Used over time, the photographer can start to see, understand and previsualize outcomes by experimenting with a wider set of compositional choices.

Photographer Stephen Shore discusses his own photographic process:

> Photography is inherently an analytic discipline. Where a painter starts with a blank canvas and builds a picture, a photographer starts with the messiness of the world and selects a picture. A photographer standing before houses and streets and people and trees and artifacts of a culture imposes an order on the scene – simplifies the jumble by giving it structure. He or she imposes this order by choosing a vantage point, choosing a frame, choosing a moment of exposure, and by selecting a plane of focus.[1]

This set of methods can be tricky though. An analytic practice utilizes, as we see here, a set of explicit steps – repeatable in any situation. I have observed there is a tendency to make this kind of process mechanical and unconscious. *I placed the focal point in all 9 quadrants of the frame and now I am done. Check!* Be aware of this. For the method to have an effect on the photographic process, engagement while photographing and thoughtful looking at the resulting photographic sets need to be active. I liken this kind of practice, since it is truly practice, to shooting free throws or practicing scales – they can either become automatic and unconscious or focused and a place to figure things out.

NOTE

1 Stephen Shore, *The Nature of Photographs* (Baltimore: The Johns Hopkins University Press, 1998), 37.

Method 8 Changing Point of View

Imagining and Embodying

Helpful with: Creating photographic images different than your normal point of view. Gaining a different experience of a place through embodying someone or something else.

Prep: Gathering props to help embody the point of view. Researching another human's or animal's habits in the landscape photographed.

Post-production steps: Attaching a written or spoken narrative to the photographic set.

Other materials: Props for changing position in the field, a notebook. Perhaps a video recorder that can also take stills or a gimbal stabilizer. Sometimes it is easier to do this method with someone else. See the following section for examples.

DISCUSSION

This method asks the photographer to physically navigate a space while imagining and embodying another person's, animal's, plant's or system's experience. Physically changing one's point of view for an entire photographic session not only results in photographs with new views but also adds another layer of understanding to that place.

Landscape architects design for many living things and systems. We design for all kinds of different humans, especially if our work is in the public realm. We design for other animals, such as insect pollinators and birds, and we design for multiple systems, from salt marshes to stormwater conveyance (see Figures 1–2). The visual experiences of these "clients" are unique. Physically photographing from their positions by imagining and then mimicking their movements through space helps the photographer/designer gain new perspectives – visually and psychologically.

This method intends to get you out of your own physical and psychological comfort zone by asking you to project yourself into another way of experiencing the world. Physically engaging in this process is key. By changing physical position, psychological experience is also changed. The photographer/designer places themselves in "someone else's shoes," or another animal's paws, wings or talons, as a way to cultivate empathy. Often, we read about other animals' experiences through discussions of larger abstract problems, such as bird strike deaths in urban areas.[1] Yet attempting to physically mimic the experience creates another level of understanding entirely.

For *Landscape Architecture Magazine*'s story about bird strikes, the cover photo of the February 2020 issue presented a bird's eye view of the situation – not from above but on the ground looking at another hummingbird lying dead in an expansion

DOI: 10.4324/9780429284649-11

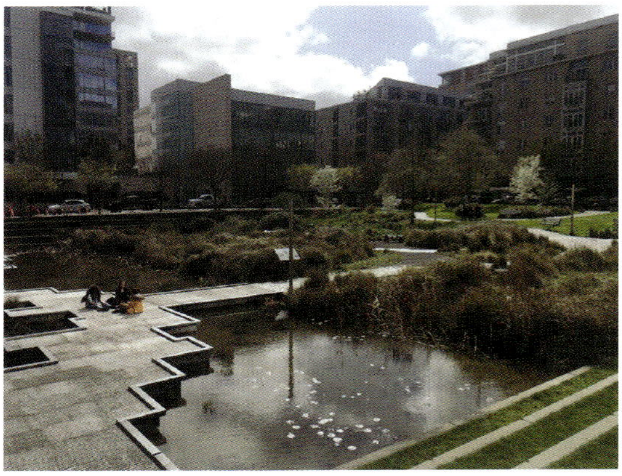

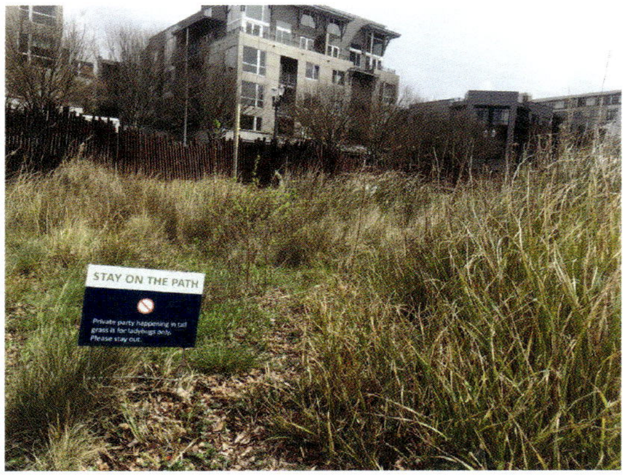

Figures 1–2
Tanner Springs Park in Portland, Oregon, is designed for many users, from humans to birds to insects. It handles localized stormwater flow and creates habitat for several animal species. The Friends of Tanner Springs post signs such as the one in Figure 2 to encourage the human users to respect the other animal users of the park.

Images: Anne C Godfrey, April 2019.

joint. The concrete fills most of the frame because the photographer's lens is almost as low to the ground as a bird's eye might be. Within the article appears another photograph that attempts to mimic how a bird would see both the reflection of trees in the glass of the atrium as well as the trees planted inside it. The photograph helps a human understand how visually confusing this situation is. It's hard to tell what is inside and what is outside.

As for understanding systems, projecting ourselves into the physical processes of systems can help us better understand some of the dynamics of that system – such as swimming in a river for several hours to count salmon.[2] So they could submerge themselves in the California portion of the Smith River for an entire day to do the count, former students of mine Bryn Davis, Keegan Oneal and Colin Poranski donned wetsuits, snorkels and flippers. Before suiting up, they were trained in methods for counting the fish, such as how to identify the females and males and differentiate between juveniles and adults. They were also instructed on how to identify the types of structures and flows in the river that were most likely to attract salmon.

In discussing this experience, all three expressed how being in the river with the salmon for such a long time radically changed their understanding of both the species and the river ecology that supports their anadromous cycle. For example, they began to "think like salmon" in order to look in the right places to count them. Of course, their experience was memorable and exciting beyond the utility of the counting itself. They engaged in an immersive methodological process (counting the fish) that transformed the way they think about systems (rivers and salmon life cycles; see Figures 3–4).[3]

METHOD

You will pick another human, animal or system to emulate while photographing. This method asks the photographer to consistently engage in the physical position of their new persona throughout the entire process of photographing. Holding the camera at

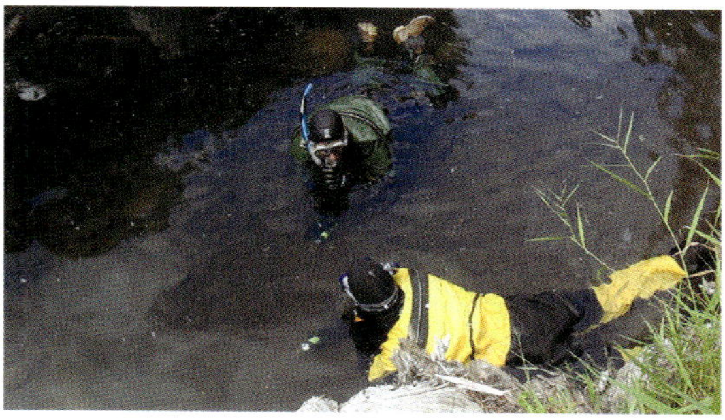

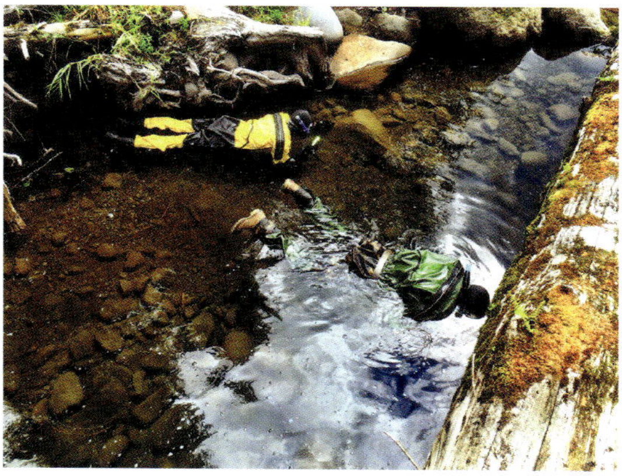

Figures 3–4
Fish counts are conducted on the Salmon River with similar methods as those conducted on the Smith River. Participants wear wetsuits or drysuits and spend the day fully immersed in the stream with the fish.

Images: Bureau of Land Management Oregon and Washington.

face level for this process also helps one enter the psychological mindset of the visual experience.

Physically changing point of view often requires the use of props. These are usually more accessible than a snorkel and wetsuit! For instance, the photographer can carry a 5-gallon plastic bucket or ladder to stand on for each photograph. As in previous methods, the photographer can entirely change their body position by kneeling, lying on the ground or crouching.

If this feels physically awkward or creates a feeling of self-consciousness (it probably will), I recommend doing this with someone else. That person can help you get into your position or help carry props. Also, when done in a small group, it might not look or feel as out of place. Several students have remarked how having a friend along really helps them get into the process.[4]

This can be taken further by psychologically emulating your persona throughout the photographing process. For example, you could start imagining and thinking like a 5-year-old child, a squirrel, a border collie, an acorn woodpecker, a landscape maintenance staff person, an ant, a bee, a blue jay, a rattlesnake, an urban horseback rider, a bike commuter or a 5′2″ 84-year-old woman. Research into the habits and experiences of this persona can add to the process, helping the photographer look, see and recognize how that other person or animal's interests may direct them in the landscape.

Over time, my students introduced the idea of emulating systems. They have photographed as water in a stream, pollen, fall leaves, a plastic bag and rain (see Figures 5–8). This is more abstract, obviously, yet still involves using imagination to project oneself in the actions of that system. Preparing for photographing turns into an opportunity to do more research about the system, figure out what needs to happen to convincingly embody the actions in the system and logistically execute the photography.

There can be no 1 to 1 emulation of the chosen animal, person or system. It is not possible to exactly replicate the visual experience of another animal through the use of the camera. For example, dogs don't see in the same color spectrum as humans. Bees and other insects see with multiple lenses and translate visual information in different ways than

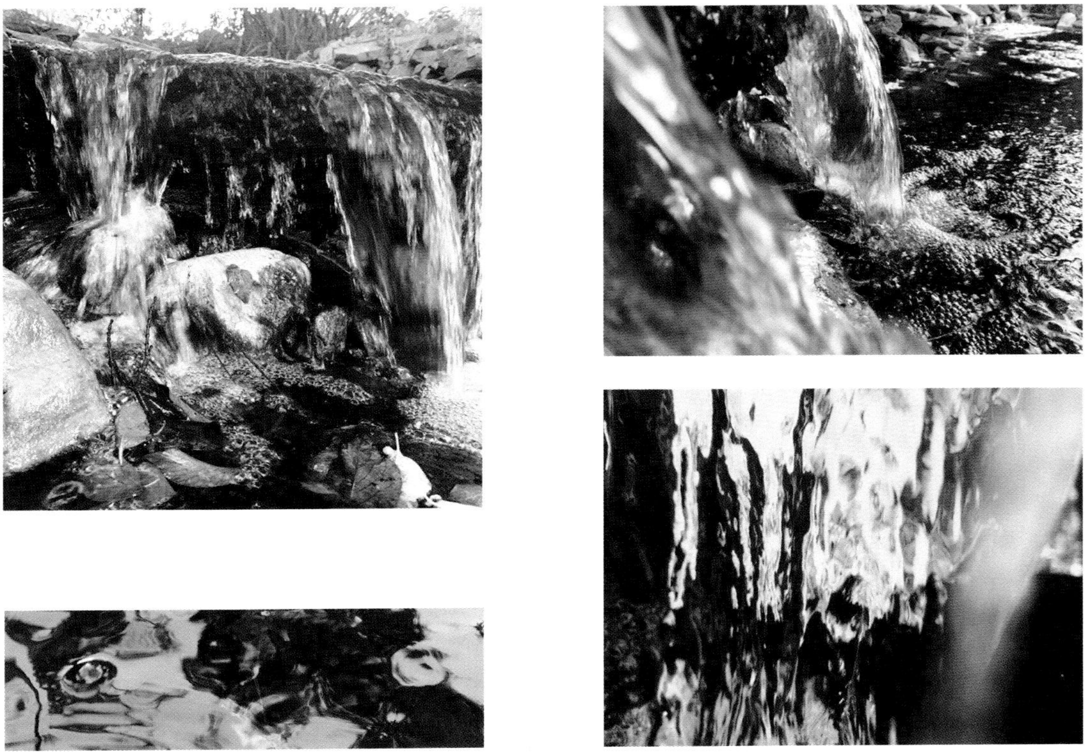

Figures 5–8
For this method, Ry Smith embodied the process of a waterfall. She explains,

"Photo 2 is her side view. Photo 3 is her breathing (bubbles). Photo 4 is through her eyes (taken behind the falls). This personifies the waterfall, making it more susceptible to our empathy. Humans tend to empathize with similar organisms, which is perhaps one reason why so many people care so little [about water]."

Images: Ry Smith, September 2019.

humans. The eye positions of birds create a different kind of cone of vision than what is represented by the camera. Systems and plants don't "see" with eyes.

Certainly, there are digital tools that we could use to emulate and simulate some of these visual experiences, but they would be produced in a lab. For this method, the process of physically photographing out in the landscape is very important. This method, to make it easily practicable, is not interested in recreating the exact visual experience of these other animals. Instead, it is about creating a process of photographing that gets us out of ourselves.

Given as an assignment, I ask students to write a short narrative emulating the person's, animal's, or system's point of view. Writing, like photographing, is a process that creates new knowledge because it is an act of synthesis. Writing such narratives has increased the empathetic experience of the students and has more deeply allowed them to project themselves into the psychological space of the physical point of view.

This is a creative act. The photographer is imagining. Through that imagining, there is an opportunity for understanding and empathy.

CONCLUSION

Over the years, students have created a surprising diversity of subjects and points of view. Here are a few examples:

An elderly woman with blurred vision (the student imagined having blurred vision while photographing, and the blur was created in Photoshop after photographing).
A bee searching for food in an urban environment.
A child in a playground.
A mother walking home with groceries and two small children in tow (this experience became about barriers and areas difficult to navigate).
A person using a wheelchair and attempting to get into the building I currently teach in (which turned out to be extremely difficult).
Rainwater being conveyed through a bioswale system (made with a waterproof go pro).
A dog on a walk in a forested park.
Trash blowing in the wind and collecting in specific places.

This method gives license for creativity and imagination, producing some very unique and interesting photographs. This process gives new meaning to the photographer's understanding of a landscape place and cultivates empathy for another living thing or system.

NOTES

1 Jeff Link, "Lethal Glass Landscapes," *Landscape Architecture Magazine*, February 2020, https://landscapearchitecturemagazine.org/2020/02/13/lethal-glass-landscapes/.
2 Thomas B. Dunklin, *Smith River Fish Count Instructional Video*, Smith River Alliance, 2009, video, 13:28, https://vimeo.com/5972885.
3 Colin Poranski, Bryn Davis, and Keegan Oneal in group discussion with the author, September 28, 2020.
4 I have been teaching this method for more than a decade. From the very beginning, I encouraged students to do this method in small groups. It helps them get over a hesitancy to really engage in this assignment.

Method 9 Photographing What Someone Else Sees

Helpful with: Understanding subject matter may be composed in many ways and interests in subject matter can vary widely from person to person.
Prep: Making time to photograph with another person.
Post-production steps: Narratives may be created detailing the process of making photographs with another person.
Other materials: A notebook or recording device to document conversations had while photographing.

DISCUSSION AND METHOD

This method is another version of changing point of view. For this method, you will pair with another person and photograph what your partner would photograph. They, in turn, will photograph what you would photograph.

How does this work?

This method asks you to describe to your partner how you would make a photograph of a particular subject matter.[1] Then your partner makes "your" photograph and shows it to you. Each of you takes turns making the other person's photograph as you traverse a landscape together (see Figure 1).

The first challenge is photographing what the other person describes. Each person must closely listen to the specific way their partner describes how they would make a photograph. The other challenge is describing a composition to another person. Each photographer practices articulating their own thoughts and interests using specific descriptive words. This highlights how the act of photographing is an act of making, selecting and creating. This also becomes another form of not photographing, in this case describing to someone else instead of making the photograph yourself.

Once your partner makes your photograph based on your description, look at it together and describe what worked and what was different than you intended. This can lead to making a second photograph to get closer to your photographic vision. Partners can either take turns, doing this photograph by photograph, or one person goes first for a designated number of photographs (not too many!), and then the other person takes their turn.

This back-and-forth dialogue often leads to describing *why*. *Why am I interested in this? What makes me want to photograph this in a particular way?* Thus, the motivations of the photographer are immediately revealed and articulated, instead of remaining

DOI: 10.4324/9780429284649-12

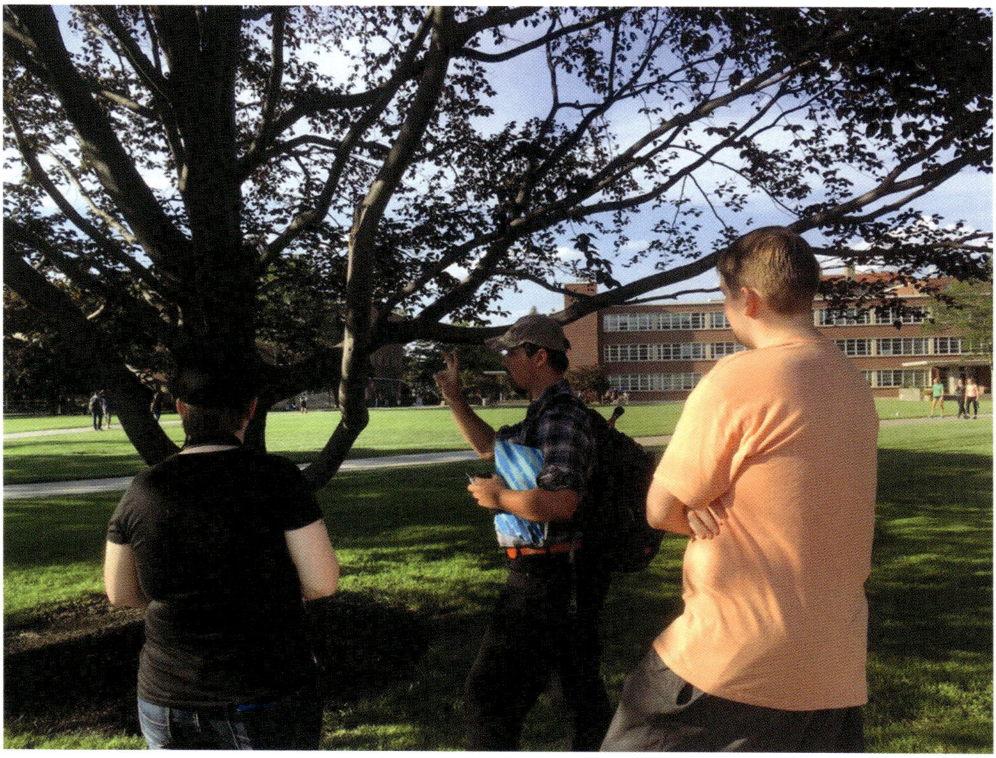

Figure 1
Here a team of three people take turns describing their photographs to one another. Often the process of describing includes hand gestures and pointing, or tracing the subject in the air. Teammates will sometimes line up behind the person describing to get into their general cone of vision.

Image: Anne C Godfrey, August 2019.

behind the image, unspoken, and unknown. This process often becomes a conversation between the two photographers, instead of simply relaying instructions (see Figure 2).

This method grew out of searching for an effective and immediate way to help students understand how differently someone else can interpret the world through photography. It also has turned into an opportunity to practice describing compositions.

Students quickly grasp that this process is conceptually complicated. Skills build as each of the participants continues to figure out how to describe their photograph to the other person, and the other person figures out how to interpret those descriptions. It creates an interesting feedback loop. This challenge is time intensive and requires a good deal of concentration. Limiting the number of photographs – such as 12–24 (with a retake making 24–48) – helps each photographer to really listen.

Another version of this method engages a stakeholder, community member or client. This iteration should also involve both people photographing. This can create a dynamic interplay of visual and nonvisual understanding of a place, values and interests through the conversations that inevitably unfold while making each other's photographs. Another outcome may be increased trust through the listening relationship that forms during this

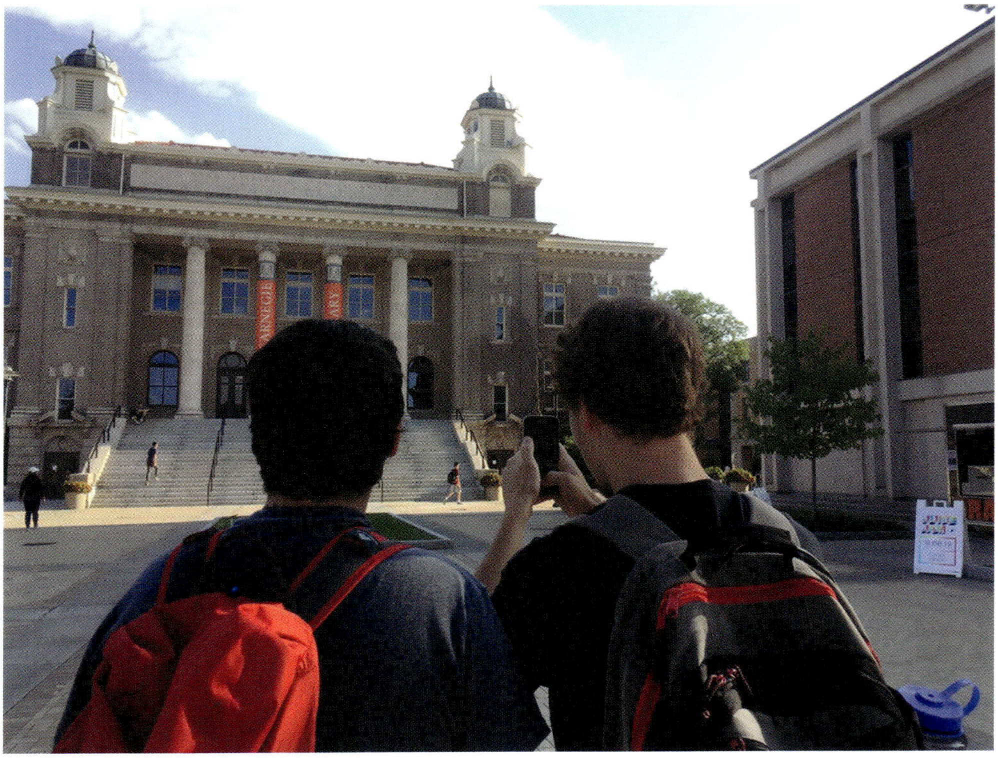

Figure 2
As the method progresses and the team members get to know one another, they each learn more about how their partner chooses to describe their photographs. It then becomes easier to work well together and successfully make each other's photographs.

Image: Anne C Godfrey, August 2019.

process. This method inherently acknowledges the validity of each of the photographer's intentions and interests. One person's photography is no better than another's; it is simply different.

This is yet another example of how a specific photographic method yields far more than the photographs themselves. This method, when done thoughtfully and respectfully, places each photographer on equal ground. In doing so, conversations can evolve in unscripted ways, allowing for a space of sharing that may not be possible in a meeting, or a lecture or a presentation. Empathy can be cultivated.

Limits

The real limit is yourself. If you can let go of your attachment to how you photograph, and truly learn from and respect the way another person photographs, this method becomes quite powerful. It requires close listening and letting go of control. Each photographer is on equal footing. If this is hard to do, think about why. Are you catching yourself "improving" upon the composition your partner is telling you to make? What does that mean? Remember, these are only photographs.

CONCLUSION

The previous method (changing your point of view through projected embodiment) asks the photographer to imagine someone or something else's experience. With this method, the other person is right there in front of you sharing what they see and think about in real time. The immediacy and reality of this situation can be quite striking.

Most of my students really enjoy this method once they give it a try. But, at first, I recognize this might feel uncomfortable. It's ok if you feel awkward doing this with a new acquaintance. There are no "right" or "better" compositions for this method. This isn't about making the best photographs. It's about how *you* make photographs. Over time, I've noticed this method help students understand how differently someone else may see the world. And often it's an interesting way to get to know someone new or learn new things about someone you already know.

NOTE

1 As defined in the introduction, subject matter could be a specific thing, or a set of things or a larger landscape scene. In many instances, it will contain all three of these types of subject matter.

Method 10 Casual Composition

Note-Taking

Helpful with: Making photographs as a way to take notes. Embracing the idea that it's ok to make photographs that "don't look good." Considering how photographs assist with remembering ideas and concepts.

DISCUSSION

This method may seem antithetical to some of the previous methods, which discuss how compositional choices made by the photographer influence how a subject matter is presented. Instead, this method addresses an assumption about photography – that useful photographs should always "look good." On the heels of this assumption is the idea that a photograph has little value if it does not look good. In many settings, from the classroom, to the studio to the office, someone will point out "we can't show that photograph because it doesn't look good."

Here, as with the previous method, Someone Else, let's take steps toward consciously allowing for more types of photographs. In this case, photographs that do not emulate aesthetic ideals or conventions.

What I term "casual" photography can be just as useful (and at times more useful) than photographs carefully following codified compositional and aesthetic rules. In this case, photographs can act as a kind of note-taking. Notes are often a short string of words, a single word or a list. Notes are not expected to be carefully constructed complete sentences. Notes are made primarily for the note-taker. Notes are often made to help remember something.

Similarly, photographic notes, made for the photographer's own purposes, don't need to follow formulaic visual conventions. Some of these conventions include the culturally derived ideal of the picturesque, the application of the mathematic rule of the golden section and the presentation of unity or balance, to name a few.[1] Within these common conventions, legibility and strong forms are often favored, as well as visual organization and clarity, or singularity.

Influences

This method of casual note-taking is informed by three situations I have encountered in my research:

1) Several practitioners brought up their personal practice of "photo note-taking" during the interview sessions I conducted for the first book in this series, *Active Landscape Photography: Theoretical Groundwork for Landscape Architecture*.

DOI: 10.4324/9780429284649-13

2) The growing acknowledgment that many standard aesthetic styles (which in the past have been treated as universal) are actually expressions of specific cultural and historical value systems from specific parts of the world.
3) Photography is an accessible tool most any person can use to describe a landscape place.

All of these situations are influenced to a greater or lesser degree by the idea that a photograph is a representation of what the photographer likes, is interested in, is concerned about, cares about, and generally thinks is important. Also, the photograph expresses what the photographer has been exposed to in the past, be that an aesthetic style, phenomena, a group of people, or a type of landscape. A diversity of different experiences and ideas can be expressed through photography if we allow for it.

From a note-taking approach, I may make a "bunch" of photographs of things. These things may not show any context and perhaps are made without any sequence. They act as a collection of things I think are important. They could be about textures, colors, geologic phenomena, postindustrial detritus, traces of animal use. It is as if I am picking these things up and putting them in my bag to reexamine later. Thus, they are like visual field notes, my pad and pen replaced by my eyes and the camera (see Figures 1–4). I am collecting. And I am not laboring over making "good-looking" photographs of these things. That is not the point.

We are used to a set of aesthetic standards applied to photographs (beauty, unity, balance, picturesque, rule of thirds, etc.) that can lead to quickly (unconsciously) discounting the usefulness or validity of photographs that do not meet these standards. I'd like to point out how this can lead to missing things and leaving multiple experiences out. Perhaps unknowingly, a rigidity develops around what kinds of photographs are considered useful or even possible. Photographer Stephen Shore discusses this in his book *The Nature of Photographs*:

> When photographers take pictures, they hold mental models in their minds; Models that are the result of the prodding's of insight, conditioning, and comprehension of the world.
>
> At one extreme, the model is rigid and ossified, bound by an accumulation of its conditioning: a photographer recognizes only subjects that fit the model, or structures pictures only in accordance with the model. A rudimentary example of this is a mental filter that permits only sunsets to pass through. At the other extreme, the model is supple and fluid, readily accommodating and adjusting to new perceptions.[2]

Shore thoughts can help us examine our own photography processes. Am I making the same kind of photograph over and over? Do I not make a photograph because I can't make it fit into my (rigid) set of aesthetic expectations? Is a kind of sorting occurring – *looks good in a photograph, does not look good in a photograph*. This situation leads us to consider: am I leaving things out, just because I can't make them "look good?"

At this point in the discussion, one is tempted to blame the camera, blame all of photography for this – photography is the reason why I don't think something looks good as a photograph. Certainly, photography has its own set of constraints – rectilinear frame, aspect ratios, cone of vision, depth of field – but so does any representational medium.

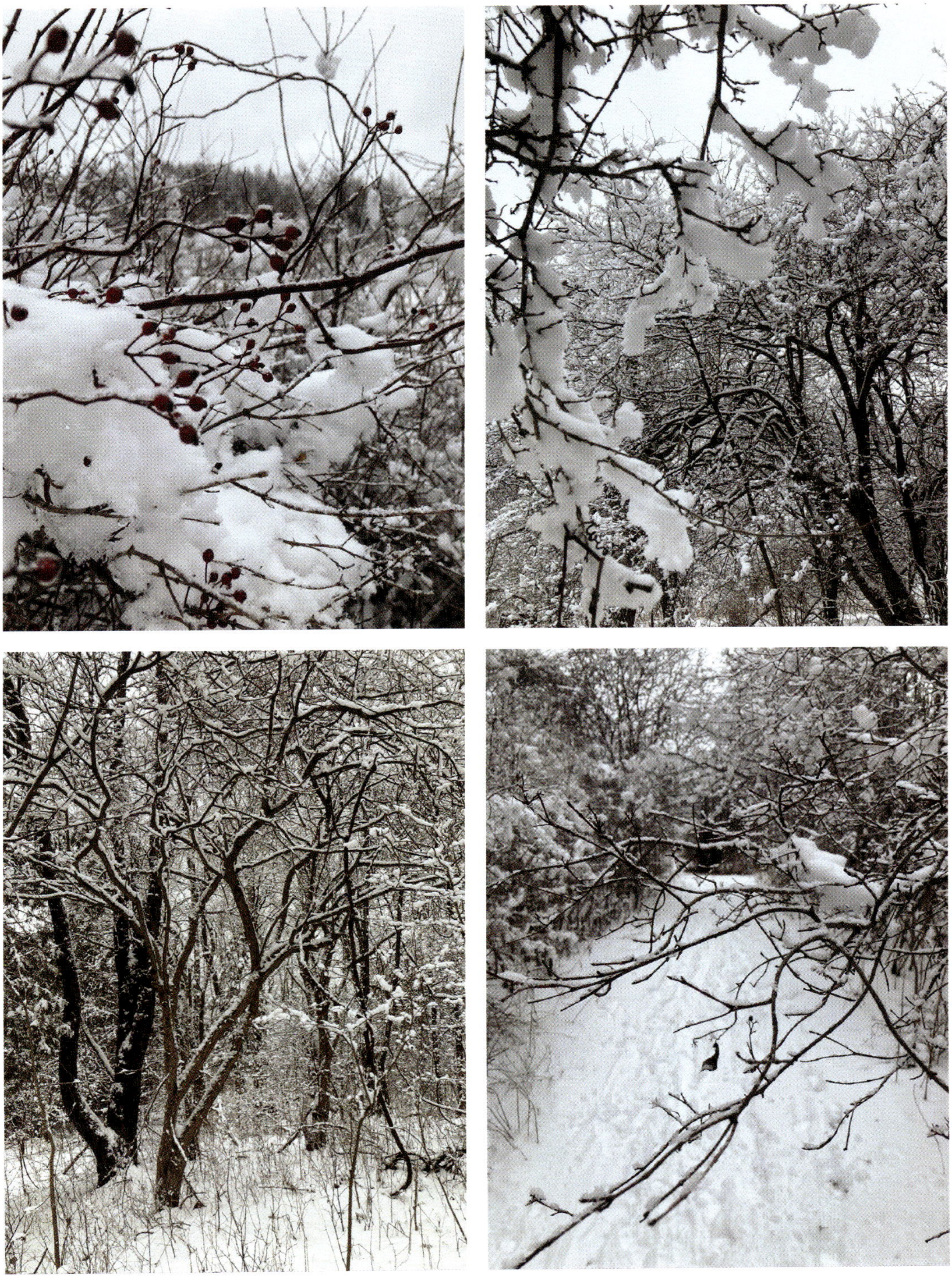

Figures 1–4
Making these photos during a wintery walk was a way to take notes about my interests in the branching patterns of the trees. Looking at the photos I know they do not visually represent what I was seeing (the depth of field is different, and the contrast is sharper), but they do remind me of my experiences. In this case, I made the photographs to explore a drawing project. The photographs are meaningful to me, but perhaps would be dismissed by someone else as not "good-looking," or more specifically, too messy and unorganized.

Images: Anne C Godfrey, January 2021.

Those constraints do not outright determine what is made. It still is a matter of the photographer exploring and trying different options available (which are far more than many think) within this form of representation.

This discussion can also be applied to how we chose photographs. This same kind of sorting – good photograph, bad photograph – occurs when assessing the value of the photographs we make and find for research and design decision making – again, often unconsciously utilizing culturally conditioned aesthetic preferences to make these choices. What could we be missing during this process?

What if we allowed for the condition that many photographs may not meet codified aesthetics, *yet* there is much to be learned from them. We could practice seeing this way through our own experimentation with photographic note-taking, allowing ourselves to make photographs that help us remember but may not look good. This could grow into taking a more open approach to looking at photographs. The photographs could be assessed on their own terms.

THE METHOD

This is both the simplest and hardest method in this set.

Photograph in an open way. Again, this sounds totally antithetical to everything presented thus far. But let's try it. What if you untied yourself from various expectations? When photographing, try to put these tendencies in check:

Making a quick assessment of what makes a good photograph or bad photograph.
Considering how the photograph could be posted on social media.
Impressing others.
Imagining it being printed in a monograph.

What if you simply photographed what you thought was interesting, important, meaningful or of concern and did not stop yourself because it wasn't a "good photograph?" What might show up in these photographs, these notes to self?

This still isn't about making a lot of photographs. Each photograph can still be made in a thoughtful way. With this note-taking approach, the 5 second rule and question asking can help the photographer look more closely and understand the reason for making the photographic note. The 5 seconds of looking could be spent seeing more details instead of figuring out the "best" composition. Looking without photographing can become a form of synthesis, combining observed experience with photographic notes.

NOTES

1 One contemporary resource: Berys Gaut and Dominic McIver Lopes, *The Routledge Companion to Aesthetics* (London and New York: Routledge, 2001).
2 Stephen Shore, *The Nature of Photographs* (Baltimore: The Johns Hopkins University Press, 1998), 117.

Method 11 Time With Photographs

Helpful with: Realizing that what is seen and understood in photographs changes as a project progresses. Getting into the habit of thoughtfully returning to and reassessing photographs used for design.

DISCUSSION

Humans possess an incredible ability to quickly look at photographs and remember their basic content.[1] Yet spending time with photographs and returning to them over and over cultivates a more dynamic relationship of understanding. What is seen and understood in photographs during the first month of a project will change, sometimes radically, as a project progresses over months and years.

This transformation of understanding is informed by the relationship of photographic meaning (photographer + photograph + viewer + context = meaning) presented in the introduction.[2] How the viewer understands the photograph is changed by the "external context"[3] of a project and how that project changes and evolves over time. Every design project requires researching new things, engaging new people and encountering new situations. This process of learning augments the way photographs are understood – something within a photograph becomes more defined once specific information is learned (see Figures 1–2).

Photographer Roselee Goldberg puts it this way:

> The photograph for me is like picking up a shard of poetry in Pompeii and finding the story of an entire civilization in a handful of material. The photographer is the first tier of the history of performance, the first line of record, and we need to learn how to read the photograph – as good art historians or archaeologists do – for its iconography, for its clues about the times, the politics, the communities in which the work took place, where it occurred.[4]

As landscape architects and environmental designers, we also read for clues. Our ability to see and interpret some of those clues evolves as the context of the project changes. We gain more information about communities, budgets, stakeholders, infrastructural constraints and history, just to name a few characteristics. The conceptual context of the photographs – as Barrett states "the situation in which the photograph is presented or found"[5] – changes as a project progresses (see Figures 3–4). Therefore, it is helpful to continue to spend

DOI: 10.4324/9780429284649-14

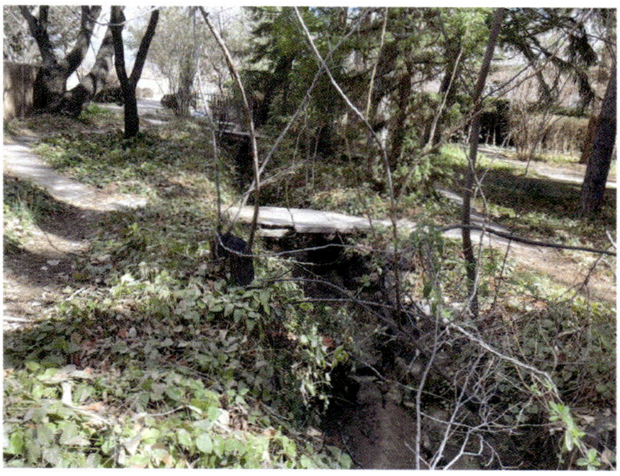
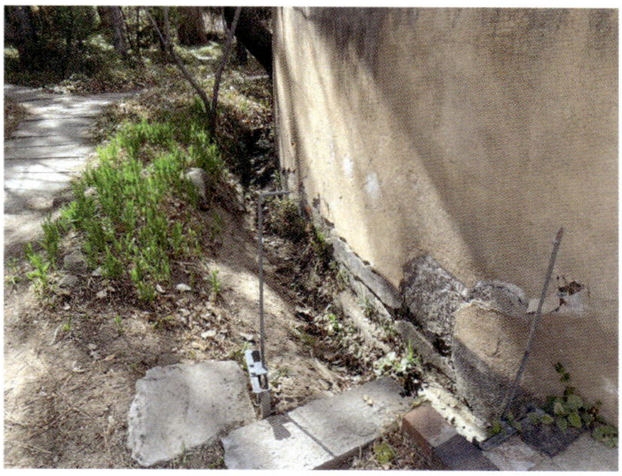

Figures 1–2
If a viewer is not familiar with this type of water conveyance system found in New Mexico, the characteristics in these photographs can be difficult to understand. After learning more about the acequia system and how it conveys irrigation water to private land, the characteristics start to become more legible. Here, the first photo is of a ditch that takes water from the main ditch onto the private property at Los Poblanos, a restored farm and hotel in Los Ranchos de Albuquerque, New Mexico. Several small gates control water flow to different parts of the property. Figure 2 shows a small gate that conveys water to the Rose Greely Garden. The water flow to this garden is primarily for the irrigation of the ornamental plants.

Images: Anne C Godfrey, March 2021.

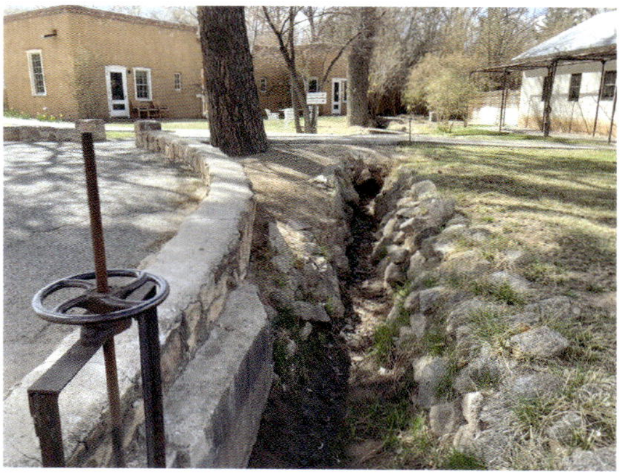
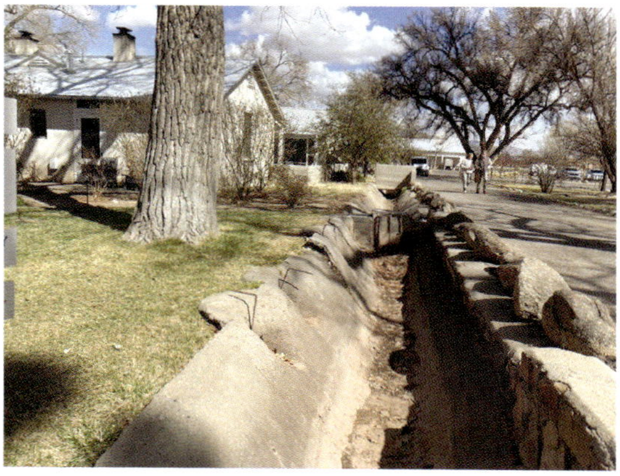

Figures 3–4
Other ditches convey water to the fields. Their gates are larger because of the higher water volume – enough to flood irrigate large fields. The condition of each of the ditches varies. Some are bare earth, while others are lined with stones or some type of hard surface, such as concrete. The ditches need to be maintained every year, either re-dug or cleared of debris. Figure 3 shows one of the main buildings of the estate, designed by John Gaw Meem. Figure 4 looks toward the main fields along Rio Grande Boulevard and the Sandia Mountains in the distance. With each successive visit, I am able to identify and understand more details about both the irrigation system and the larger cultural landscape at Los Poblanos.

Images: Anne C Godfrey, March 2021.

time with the photographs, returning to them over and over to reassess what we see and learn from them.

Often, photographs are consulted at particular stages in the design process and then left to the side. A subset may remain pinned up at desks, on in-process project walls or other kinds of display areas. But more likely they remain in digital folders accessed through a computer. How often are these photographs revisited? Are they looked at less and less as the project continues? As the context and knowledge of the project changes, those photographs can still yield interesting new information.

THE METHOD

This method calls for pointedly returning to and reassessing photographs at multiple stages throughout the design process. Returning to photographs at predetermined stages throughout design development becomes a way to check in. *Is what I am remembering aligning with what is presented in this set of photographs? Or not? I keep remembering a particular feature. Why? Let me look at these photographs again to check. Oh, I didn't see that last time. Now I think about this differently.*

I might remember certain features from when I first looked at the photographs that I identified as having the most importance. In this case, the context of the initial stages of a project points us toward one set of fundamental interests. While the context of a more advanced stage of design implementation may compel me to identify other details or relationships as important.

In certain cases, spending time with photographs can be intertwined with site visits at more progressed stages in the process. Returning to and looking at those initial photographs in comparison to more recent site visit photos is a way to check in. Photographs made at the early stages of a project can be brought and viewed on-site during a later visit as another method for checking in. In other cases, with remote site design, the initial photographs remain the primary source of visual information about the site and its context. Critically assessing them later in a project can still yield new knowledge. This demonstrates how an active ongoing relationship occurs between viewer and photograph. Certain questions facilitate this method of checking in:

> What did I initially find important in these photographs?
> Is that the same now?
> What new things do I see and understand today that were not within my realm of understanding when we started this project?

Written narratives can be attached to photographs, a technique introduced in many of the previous methods. These narratives can then be consulted during the regularly scheduled process of returning, looking and reflecting. Adding new narratives tracks progress and change.

CONCLUSION

Spending time with photographs, as with many of these methods, doesn't have to take too much time. Taking some time to look again makes space for the designer to find new information in the photographs. At this point in the progression of these methods, this more sophisticated and critical understanding of how photographs work is put into practice – the context and meaning of photographs change over time.

NOTES

1 Jeremy Wolfe, "Photography Changes What and How Much We Remember," in *Photography Changes Everything*, ed. Marvin Heiferman (New York: Aperture, 2012).
2 Also presented in Anne C Godfrey, "Meaning Is a Relationship: Gathering and Photographs to Create Landscape Knowledge," in *Active Landscape Photography: Theoretical Groundwork for Landscape Architecture* (London: Routledge, 2020), 48–60.
3 Terry Barrett presents this concept in his widely utilized text *Criticizing Photographs: An Introduction to Understanding Images*. Terry Barrett, *Criticizing Photographs: An Introduction to Understanding Images* (Boston, MA: McGraw Hill Publishers, 2000 3rd edition), 96–103. I re-present this concept for uses in landscape architecture in Ibid *Active Landscape Photography: Theoretical Groundwork for Landscape Architecture*, 51–53.
4 Melissa Harris and Michael Famighetti, "RoseLee Goldberg and Roxana Marcoci," in *Aperture Conversations: 1985 to the Present*, eds. Melissa Harris and Michael Famighetti (2015; New York: Aperture, 2018), 178.
5 Barrett, *Criticizing Photographs*, 99.

Method 12 Significance of the Unphotographed

DISCUSSION

During the design process, a vast collection of photographs (made by ourselves; shared by the client, collaborators and stakeholders; gathered from historic archives and government sources; and mined from the internet) are consulted. These collections present both interrelated and incongruent visual information.

One example of incongruity becomes clear when we start to notice what is not photographed. This method focuses on considering what is not present in the photographic documentation of a place. This method also presents the next step: asking *why* particular characteristics or relationships are not photographed.

Photography is collected to gain a better understanding of any site and context for which we design. This creates an archive, made through the process of choosing – deciding what is important, what is less important and not important. We simultaneously choose and not choose. As discussed throughout this book, what is not photographed can reveal equally important issues as what is photographed.

Two specific situations occur within the condition of the not photographed:

1) Photographs of certain subjects or relationships are not even present in the available collections of photographs we choose from.
2) Particular photographs of subjects or relationships are consistently not chosen, even though they exist in collections of available photographs.

THE METHOD

Situation One: What Is Missing?

If something is not photographed, it does not become historically relevant or maintain a presence in the conversation because there are no photographs. This concept is discussed in multiple disciplines, such as contemporary practices in history and media culture. The idea of absence is often utilized when interrogating power structures – those in power create the visual narrative (as well as the textual narrative), while another culture's narrative is diminished or erased. This is a foundational theme in postcolonial studies.[1]

For example, not seeing certain kinds of bodies or races or genders in popular entertainment diminishes the validity and realness of those people. If women over 50 continue to not be cast in leading roles that portray women over 50, that muddles thinking.

DOI: 10.4324/9780429284649-15

Similarly, if certain races are written out of books or comics produced for the big screen (such as Asian women being swapped out for white women[2]) their existence is essentially erased from the understanding of that work. The cycle continues – we don't see something that we don't expect because we don't expect it to exist because we don't see it.

This is true for landscape characteristics as well. As discussed in the previous methods, the aesthetic preferences we've been trained to favor through education, mass media and specialization may lead us to not choose, or in this case not photograph, certain landscape and place characteristics. There may be no photographic "document" of certain aspects of a place because we have unconsciously not photographed it, or someone else may not have photographed it because it doesn't measure up to some of these aesthetic ideals. Similarly, characteristics are not photographed because they are not culturally valued or deemed important or even necessary, and thus they are not photographed – an act of ignoring or devaluing.

Yet, landscape architects are in a unique position to assess what is not there. Our training and daily practice are interdisciplinary by nature, thus we hold, access and utilize multiple systems of knowledge. Making informed choices for design is dependent on understanding and working with interrelated characteristics and systems. Our expanding knowledge of ecological systems, for instance, allows us to understand when parts of a system may not or cannot be visually represented in photographs. In other situations, we are able to imagine specific relationships based on our knowledge and start to understand when a particular (important) relationship is not showing up (perhaps because it is "messy" or "chaotic" or was not considered valuable or meaningful to those who held power in the past). This applies to cultural systems as well. We possess the research skills and acuity to assess who is not there, yet should be, and then start understanding why.

Situation Two: Checking In

For the second situation, the goal is to check in and see if there are characteristics or relationships that are not being photographed or chosen from available photographs. Attending to this condition may include scheduling check-ins during a project to reassessing the photographic set used (similar to the previous method).

This process also keeps the photographic set, or archive, created for a project alive and continually growing and changing. The unique photographic archive made for any project is consulted over and over. Something that was always there may be seen anew. Or gaps start to become apparent that can then be filled with new photographs.

It could be something as simple as this: there are no dogs in these photographs of the neighborhood we are working for. Yet we watch many people walk their dogs. And we have heard people say in community meetings that dogs are important to them. Why don't we see this in the photographs?

Or it could be more complicated. This historic archive has no photographs of local indigenous tribe members or their homes, which were adjacent to the design site. Yet in these written documents, we read about the frequent interactions between the Anglo industrialist and the tribe members. What is going on here?

Photographer Latoya Ruby Frazier discusses this phenomenon in relationship to her own research for *The Notion of Family*, which presents photographs of her family alongside photographs documenting how her hometown Braddock, Pennsylvania, has been changed by the steel and chemical industry. Frazier discusses an encounter with a photography book during her research process:

> I purchased the book *Images of America, Braddock, Allegheny County* [2008]. I took the book to my studio, excited to learn through images about the many men and women that migrated from Scotland, Ireland, Germany, Russia, Italy, England, Sweden, Finland, Norway, the American South, and Africa. By the time I turned to the last page I realized that all of the African Americans who had contributed to this great history were excluded. This continued omission, erasure, invisibility, and the silence surrounding African American sacrifices to Braddock and the American grand narrative is why I've chosen to work with the 20th century documentary aesthetic. It is also the reason why it is necessary to tell the story of three generations whose lives paralleled the rise and fall of the steel mill industry – and to tell how we survived the subsequent 30 years of disinvestment and abandonment by local, state, and federal governments.[3]

CONCLUSION

Understanding the significance of the unphotographed is another path to acknowledging photographs as subjective forms of representation. Individual photographers carry their experiences, beliefs and biases with them while photographing, which influence the subject matter of photographs, how they are made or even *if* they are made. Photographs are forms of representation that reinforce narratives about place, history and culture.

This method highlights how photographs are freighted with all kinds of agendas and interests – sometimes understood and known, and sometimes internalized and unconscious. As landscape architects, we are trained to step back and apply our complex sets of experience, training and knowledge to ascertain the landscape situations we encounter. We can apply these skills to assessing what is not photographed, then actively discuss the implications and, finally, make choices to expand the subject matter represented with photography.

NOTES

1 For example, see Tariq Jazeel, *Postcolonialism* (London: Routledge, 2019). For a discussion of the use of photographic representation in reenforcing colonialist hegemony, see Eleanor M. Hight and Gary D. Sampson, *Colonialist Photography: Imag(in)ing Race and Place* (London: Routledge, 2002).

2 For example, Scarlet Johannsen played "Major" in the live action movie *Ghost in the Shell*, while her character in the original manga series was Japanese, named Major Motoko Kusanagi. Another example comes from the movie adaptation of *The Martian*. NASA employee Mindy Park, who was described in the novel by Andy Weir as Korean-American, was portrayed by white actress Mackenzie Davis. Amanda Hess, "Asian-American Actors Are Fighting for Visibility. They Will Not Be Ignored," *New York Times* (New York), May 25, 2016.

3 Dawoud Bey, "LaToya Ruby Frazier," in *Aperture Conversations: 1985 to the Present*, eds. Melissa Harris and Michael Famighetti (2014; New York: Aperture, 2018), 156–57.

PART II

Practices

Part II: Practices builds upon the foundations created in Part I: Methods. Here, specific examples from contemporary art photography practices are examined. The primary interests of each of the photographers is discussed, thier practices are articulated as specific methods and then extrapolated into situations encountered in landscape architecture and environmental design. The photographic methods presented here are defined by contemporary critical theories and complex working practices.

DOI: 10.4324/9780429284649-16

Practice 1 Series: Narrative

DISCUSSION

This method is interested in how a specific story about a place can be conveyed through a series of photographs. The photographs are made and presented in a way that enables the viewer to interpret a connected narrative while looking at the sequential set of images.

How can we begin to construct a process to create photographic narrative sets? Start with questions:

- What do I want to know about this place?
- What is important for others to understand about this place?
- What stories are in plain sight?
- What stories might be revealed through pointed and preplanned photographic investigations?
- Later in the design process: what foundational narratives are inspiring design decision making?

All these questions ask the photographer to cultivate a more complex understanding of the place before even photographing. Asking these questions leads to research in multiple areas. This research assists in contextualizing what is seen and experienced during site visits. A clearer relationship forms between tangible qualities on the site and invisible forces or past histories uncovered through research. The landscape is "read" as the photographer/designer gains more knowledge to help them understand what they encounter.[1] For this method, intentionally making photographic series to display these narratives is the goal. These narratives can become bright conceptual threads running through a project.

INFLUENCES

The storyboard, commonly used in filmmaking, is a method for mapping out a narrative, scene by scene (see Figure 1). Often done as drawings, like a comic book or graphic novel, the storyboard outlines the major trajectories of the film and locates significant scenes, events and pivots in the story line. These storyboards can be quite involved, with hundreds of frames, so this isn't directly transferable to the needs of our work. Yet this

DOI: 10.4324/9780429284649-17

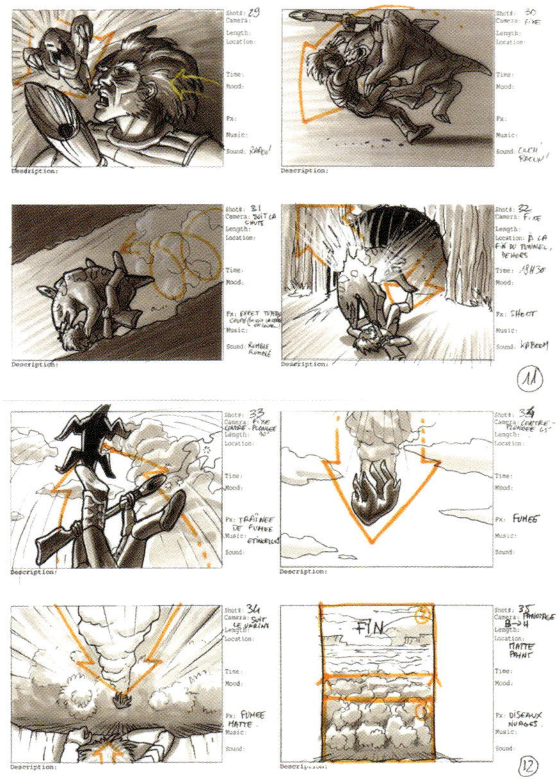

Figure 1
This is a single page of a multipage storyboard sketching out a short animated film. This part of the storyboard shows the final scenes. The orange arrows indicate direction of action.

Image: Ryzom MMORPG, creative commons.

process of identifying key structural elements in the "arc of a story" can teach us how to construct shorter legible narratives:

Establishing setting.
Introducing the main "characters."
Introduction of conflict or primary plot device.
Use of framing and composition to establish significance.
Rise in action.
Occurrence of a significant event (climax).
Resolution or lack thereof.
Ending.

For our purposes in design, each of the parts of a narrative may be established through a series of interrelated photographs. By creating a storyboard with a limited number of frames, one highlights significant "moments" that are crucial to moving the story along – to leave them out would obscure the sense of the story or make the narrative opaque. Storyboarding is a way to consider the characteristics, processes, people, and histories that are narrative keystones for a specific place.

Where do we start?

First, ask and answer the questions that help construct the backbone of a narrative. The list presented earlier, which is considered a more common story structure,[2] is a place to start, or you can establish your own structure that addresses the first set of questions presented in this chapter. Central to this process is understanding what needs to be present for the narrative to be legible to others. If something were absent, would the visual narrative lose its ability to be significantly understood?

Finally, asking *why am I telling this story?* becomes important. In narrative making, we place ourselves in the position to speak for others as well as ourselves. The responsibilities that come with this narrative sharing and making are significant.

What am I trying to accomplish here? Who gains from this? How does this relate to our values for the project? What am I leaving out? What are the implications? What biases are occurring here? Do I need to reevaluate some assumptions? Is there another story that needs to be told? During the design process, we act as a lens through which other people and other places are understood. How do we individually and as a design team address the implications of this position? Asking these kinds of questions is a way to start.

DUANE MICHALS: STORYTELLER

Duane Michals's work explores the construction of photographic narratives interested in intimate human moments and interactions. His approach presents one way to understand how a series of photographs can portray dynamic and nuanced situations.

Michals first started constructing narrative series of photographs, which he calls sequences, in the 1960s. In 1968, he presented this work publicly at the Underground Gallery in Greenwich Village, New York City. Sequences featured such narratives as *A Man Going to Heaven* (1967), *Paradise Regained* (1968) and *The Woman is Frightened by a Door* (1966). All are visual and conceptual constructions of complex narratives concerned with the human condition. Michals is interested in showing these unseeable ideas through photography.

For example, *A Man Going to Heaven* consists of a series of five photographs. The first photo is dark – the viewer picks out a figure of a man at the base of a staircase. As the sequence unfolds, the man ascends the staircase as a door opens at the top, flooding the scene with more light as the man ascends. In each of these frames, the figure and staircase become more lit. In the last two images, the man is surrounded and then almost blotted out by the hot white light coming out of the door.

This series is typical of Michals's interest in creating photographs of what you do not see, what is not "documentable." He states, "I don't walk around with a camera waiting for something to take a picture of. I never did. Essentially, I've always been introspective, and it's always been there, the notion that the mind is the source of everything. I'm an anomaly in photography because I don't believe in the eyes!"[3] This process is decidedly not reportage but instead careful construction (see Method 4 "Not Photographing").

When Michals first showed these sequences publicly, the work was met with confusion and outright derision by some because Michals was not using photography as a form of reportage. For example, street photographer Garry Winogrand walked out of the exhibit and later stated that "This isn't photography."[4] At this time, the dominant idea about photography was that it was documentary in nature – that it showed the "real" and captured a moment – thus street photography was the ideal practice. The single reportage photographic image was considered the pure and correct articulation of the nature of the photographic form.

In contrast, Michals is interested in constructing the real through the conceptual. Michals conceives of the sequences by combining his own experiences with his interest in pursuing larger life questions. He takes the story form, a combination of the real and the conceptual, and plays it through the functions of photography. Michals constructs narratives with visual photographic images that tap into invisible concepts and feelings, about death, love, longing, acceptance, rejection and discovery.[5] These constructions are predetermined. Once the scene, figures and props are set (which can take a longer amount of time than the photographing itself in some instances), he takes no more than a couple of hours to photograph the sequences. This approach prompts the viewer to add their own understanding to the reading of the story – through both their individual unique experiences and the larger existential questions they may hold.

For example, *Paradise Regained*, a sequence of six photographs, starts with a clothed man and woman in a simple apartment. In each successive photograph, plants are introduced, and pieces of clothing are removed. In the last photograph, the two figures are now naked, surrounded by plants, and the trappings of the apartment are all but disappeared. Michals shows us a passing event, but this is by no means reportage. Prompted by the title *Paradise Regained*, of course, he is relying on the viewer possessing a fundamental understanding of the trope of Adam, Eve and the Garden of Eden. He plays out the events of the trope with a set of highly suggestive props. Yet the setting, an apartment, and the characters, contemporary humans wearing contemporary clothing, place the viewer in a context that is today, not the past. This is where the tension builds as the photographs, the trope and the viewer combine to make something more than the sum of its parts.

This way of working echoes British photographer Henry Peach Robinson's photographic construction *Fading Away* (1858) and acts as a precursor to, for example, Canadian photographer Jeff Wall's tryptic *I giardini/The Gardens* (2017).[6] Robinson (1830–1901) made groundbreaking innovations with photography fairly early in the life of the technology. *Fading Away*, a photograph of a young woman on her death bed, includes four human figures (a matron, a nurse, the dying woman and a male figure looking out a window). The figures were not originally photographed together in that room. Instead, this photograph was made by combining five separate negatives with darkroom techniques. Wall presents a similar contemporary approach to photographic construction. Two "characters" appear multiple times within two of the three images of the *I giardini/The Gardens* tryptic. The characters appear to be talking with their duplicate selves. In this case, Wall constructs these multiple appearances digitally.

These two examples spanning the history of photography take a decidedly different approach than reportage. Wall famously states that he chooses not to photograph.[7] This disrupts the idea that all photography is and should be reportage or journalistic in nature. Michals's photographic practice centers on disrupting the assumption of photographic reportage as well. In an interview from 1974 with William Jenkins,[8] Michals discusses how photography in the mid-20th century United States was primarily reportage. To paraphrase: everybody was used to looking at reportage photography whether they knew it was reportage photography or not. Journalistic photography, as a near equal to reportage, was the most common form of photography people encountered. Thus, capturing a moment was considered the normal or the primary expression of photography. Everyday audiences of photography only saw this type of photography in newspapers, and magazines like *Time* and *Life*. So, they had no reference for realizing there were many types of photography and that photography could do many things.[9] The kind of constructive work Robinson had done in the 19th century was not part of the visual or conceptual vocabulary of the everyday viewer of photography in the mid-20th century.

In the interview Michals states pointedly to Jenkins,

> Most photographs reinforce all our prejudices about reality. They tell us that's exactly what an apple looks like, that's what a person looks like, that's what automobiles and American flags look like. I'm for photographs that make us question everything that's going on. Photographs that do not give you any answers. That make you wonder what the hell is happening. The photograph should never pat us on the back and say, yes that's a sunset, that's beautiful. They should make us ask, what is going on there?[10]

Michals's best-known sequences include *Chance Meeting* (1970), *Things are Queer* (1973) and *Death Comes to the Old Lady* (1969). Michals continues to expand this body of work, more recently with sequences such as *Taxidermied Duane on Permanent Display in a Shop Window* (2017).[11] These sequences explore the human condition while utilizing a clearly progressing story structure.[12]

Michals states, "In a way, I've always been at war with what the still photograph did."[13] His ongoing work with the sequences allows him to open up narratives by exploring both what is seen and unseen. Many of his photographs and sequences include writing that reveals the unseen, speaks of unseen psychological turmoil or simply the history of the event surrounding the singular moment of the photograph (see a discussion of this work in Practice 10 "Writing What You Don't See").

Stephen Perloff, editor of *The Photo Review*, explains "[Michals] showed that you didn't have to create a photograph as a single black-and-white rectangle. It opened the door for other photographers to do similar things to find their own paths."[14]

For our purposes, Michals's approach presents a way to move beyond thinking a single photograph can tell and show us everything we need to know. By creating sequences, Michals breaks the static nature of the single image, instead inviting a dialogue with a series of successive images to reveal a complex narrative. This method allows for an intricate, nuanced and richer understanding of a particular situation, place, or state of mind (see Figures 2–6). He helps us see how all photographs are constructions that combine the seeable with invisible concepts, histories and ideas.

Making sequences or narratives about sites and their contexts allows ourselves, as the makers and users of these photographs, and other viewers, be they collaborators, clients, stakeholders or interested individuals, to see more and understand more about these places because the narratives present information and ideas beyond a depiction (a form of reportage). Working this way helps us understand how photography can work in many ways, not just as a mode of reportage.

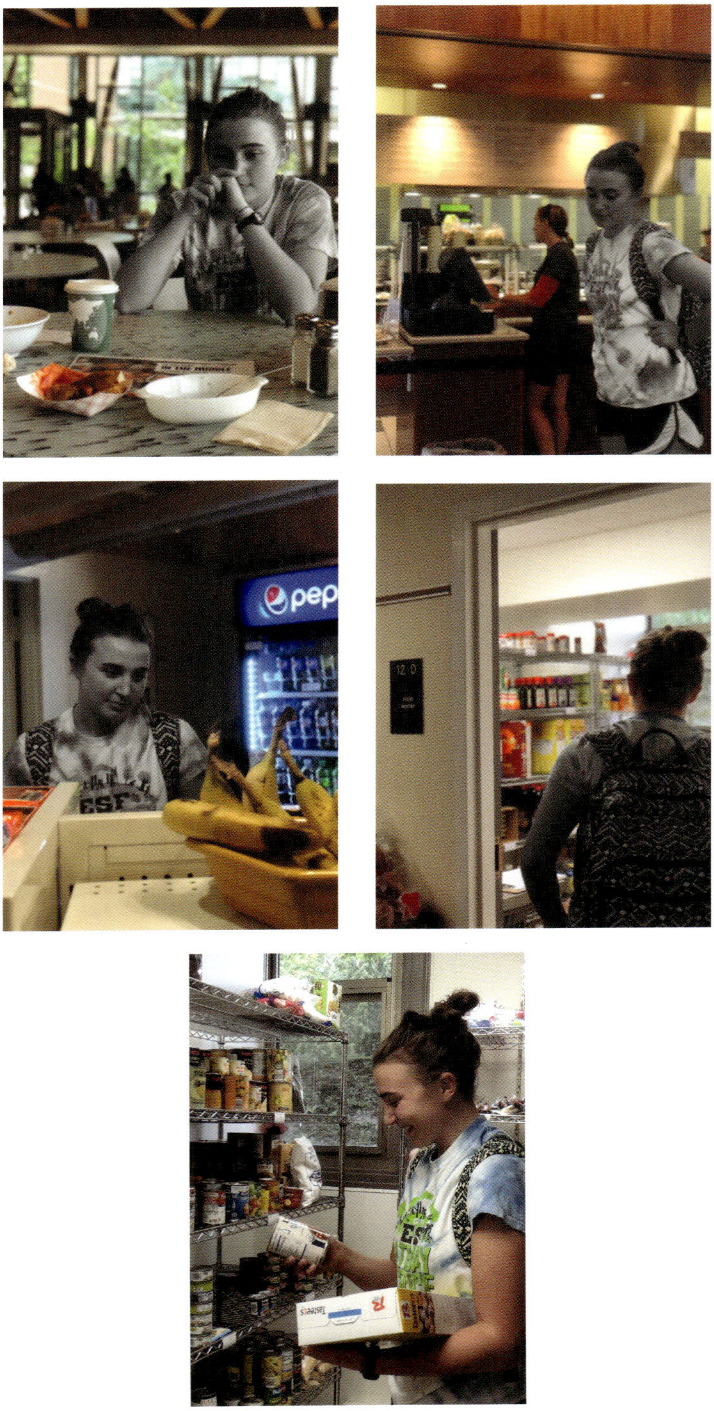

Figure 2–6
This narrative articulates a common but often unaddressed issue on campuses: student food insecurity. Here, the photographer, through this five-frame narrative, presents the main character, introduces the situation or tension and then leads us to resolution as the student arrives at the campus food pantry. The photographer also utilizes a shift from black and white to color to highlight the emotional changes occurring in the main character.

Images: Emily Plumber, October 2019.

CONCLUSION

Thus, Michals reveals multiple paths for us. One, a thoughtfully aware construction of a sequence of photographs tells multifaceted stories about a landscape. Two, we can use the arc of the story structure to assist in this storytelling. Three, these narrative sequences present both visible and invisible qualities of a place. Four, this method helps us question our reliance on single, stand-alone, reportage-like images.

Stories are usually told after the fact; narratives tell us about something that has already happened. But for our work, we are applying this process of narrative making within the process of creating something. Narratives from the past intertwine with the narratives we are unfolding in the present for the future.

NOTES

1 For example, Donald W. Meinig, ed., *The Interpretation of Ordinary Landscapes: Geographical Essays* (New York: Oxford University Press, 1979). This collection of essays establishes conceptual foundations for considering landscapes from many different positions. The importance of vernacular and everyday landscapes is laid forth. One conceptual thread articulates the notion of "reading the landscape" and accepting that there are many positions from which the landscape can be looked at and interpreted.

2 The playbook for film structure is presented in several books, such as *Save the Cat* by Blake Snyder. Joseph Campbell's *The Hero with a Thousand Faces* is widely considered the inspiration for the articulation of this story structure. More recently, Maria Tatar presents more diverse structures, paths and tools at the disposal of the main character and the progression of story in her book *The Heroine with 1001 Faces*. Blake Snyder, *Save the Cat! The Last Book On Screenwriting You'll Ever Need* (Studio City, CA: M. Wiese Productions, 2005); Joseph Campbell, *The Hero with a Thousand Faces* (New York: Meridian, 1956); Maria Tatar, *The Heroine with 1001 Faces* (New York: Liveright Publishing Corporation, 2021).

3 Linda Benedict-Jones, "Duane Michals: Storyteller," essay, in *Storyteller: The Photographs of Duane Michals* (Munich, Germany: Prestel Verlag, 2014), 33.18–57.

4 Christina Rouvalis, "Duane Michals: Telling the Story of the Storyteller – Carnegie Museum of Art," *Storyboard*, last modified October 28, 2014, https://storyboard.cmoa.org/2014/10/duane-michals-telling-the-story-of-the-storyteller/.

5 Benedict-Jones, Linda, Allen Ellenzweig, Aaron Schuman, Adam Ryan, and Marah Gubar, *Storyteller: The Photographs of Duane Michals* (Munich: Prestel Verlag, 2014).

6 In which Wall presents the same two "characters" multiple times within two of the three images of the tryptic.

7 *Jeff Wall: "Tattoos and Shadows,"* San Francisco Museum of Modern Art, March 2007, video, 2:21, www.sfmoma.org/watch/jeff-wall-i-begin-by-not-photographing/.

8 This is before Jenkins would go on to curate the groundbreaking photography exhibit at the George Eastman House "The New Topographics." in 1975.

9 Duane Michals, "This Is a Real Dream," interview by William Jenkins, in *Storyteller: The Photographs of Duane Michals* (Munich: Prestel Verlag, 2014), 84–88. Originally published in *The Photographer's Choice: A Book of Portfolios and Critical Opinions* (Danbury, NH: Addison House, 1975), 35–44.

10 Ibid., 88.

11 Duane Michals, *Mischievous Eye*, New York, DC Moore Gallery, November 14, 2019–January 4, 2020.

12 This is even the case with Michals's interests in surrealism, especially the work of Renne Magrite. For example, see Duane Michals, *ABCDuane: A Duane Michals Primer* (New York: Monacelli

Press, 2014), 54–55, 118–19. Michals dedicated his 1974 book *Album* to Margite. *Things Are Queer* is considered to be influenced by Magrite's 1952 painting *Les valeurs personnelles*. Michals states, "Everything you see is in your head." "*Illusions of the Photographer: Duane Michals at the Morgan*," The Morgan Library & Museum, last modified December 19, 2019, video, 4:19, www.themorgan.org/exhibitions/duane-michals.

13 Rebecca Robertson, "Duane Michals: Fighting Against Photography," *ARTnews.com*, July 29, 2013, www.artnews.com/art-news/artists/duane-michals-fighting-against-photography-2274/.

14 Rouvalis, "Duane Michals."

Practice 2 Predetermined Path

Regularized Photographing

DISCUSSION

Many landscape sites have strong fronts, dominating landmarks and more interesting places within the larger site. A passive photographic process focuses on these types of places. As a result, the back-of-things, or the around-over-there or that boring place doesn't get photographed. Creating a predetermined path with regularized photo points is one way to transcend this tendency. This method explores how implementing predetermined ways of navigating a site prevents the photographer/designer from primarily photographing highly legible characteristics.

Kevin Lynch's *Image of the City*, a foundational example from landscape architecture, articulates this bias toward the legible. He conducted community mapping studies with residents in the Boston, Massachusetts, area.[1] Through interviews and cognitive mapping sessions, Lynch's graduate students created maps synthesizing the trends in the residents' memories of the area. This cognitive mapping process demonstrated how certain characteristics would stand in high resolution, while whole other parts of the city were not remembered or named. Lynch comes to the conclusion that these remembered elements (which he names Paths, Edges, Districts, Nodes and Landmarks) are what make places imageable.[2] This study also displays how other, likely important, characteristics can easily be left out, primarily because of how they look. Another example, briefly here, is Guy Debord's psychogeographical dérives resulting in maps, such as *Naked City*,[3] where whole parts of Paris, in this case, are left out or articulated simply with red arrows. (See Practice 8 "Rules: Games, Drifts, Prompts.")

This method counters desirous landscape wandering influenced by highly legible characteristics through predetermining navigational paths and points of view for photographing. A few examples include the following:

- Creating transects across the site and photographing at predetermined points along the line of the transect. These transects cut across both places humans are intended and not intended to be – such as sidewalks and planter beds.
- Photographing at 90- and 45-degree angles away from or toward significant landmarks (this often results in circumnavigating the landmark).
- Photographing forward and backward, or in each of the cardinal directions, at predetermined points (see Figures 1–12).

DOI: 10.4324/9780429284649-18

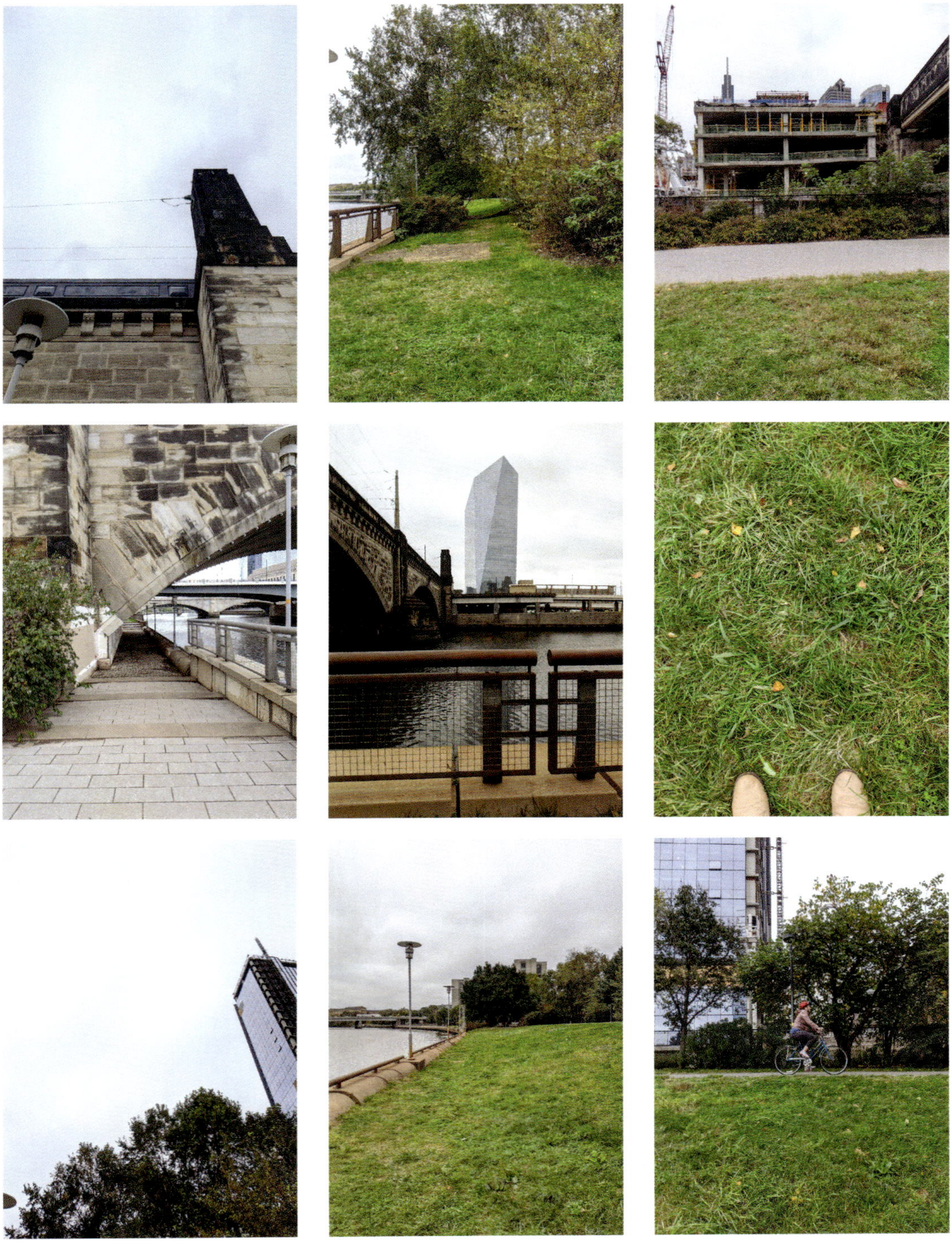

Figures 1–12
Here, the photographer stopped to make a set of photographs every 20 paces. The progression of photographs made at each stopping point looks up, then in four directions and ends with photographing the ground at her feet. This process helps the photographer look in many directions, instead of simply looking forward along the length of the path. Philadelphia, Pennsylvania. The images are read from left to right, with each set taking up two rows.

Images: Elizabeth Gargiulo, October 2020.

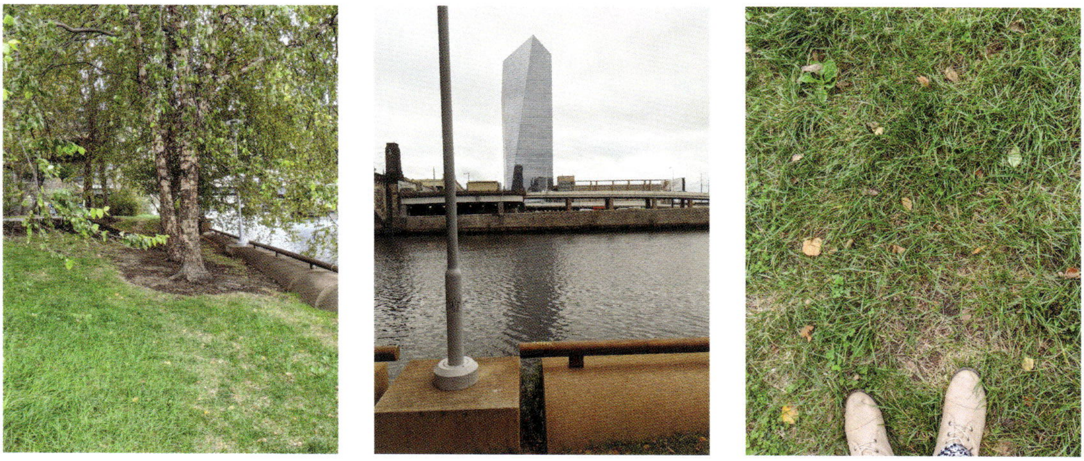

Figures 1–12
(Continued)

These regularized methods use forms of predetermined measurement (distances, angles, steps, directions) to determine what is photographed. This helps the photographer photograph more than simply what they find interesting, visually compelling or dominant. Here, "more" means that more parts and places of the site are photographed because the predetermined path and points of view cover a more regularized ground. There are fewer empty places along the way because the method directs photographing along the entire path, not just the destination or points of interest. Additionally, one can set their camera to a fixed focal length, preventing the photographer from zooming in and out at will.[4]

The person photographing navigates through parts of the site they may not otherwise because the path is prechosen, instead of chosen by the photographer in the moment. Something as simple as a muddy expanse or rough ground with trash and debris may cause the photographer to avoid moving through those areas. Anticipating less than ideal conditions helps this method become practicable. Sometimes traversing parts of the site is simply not possible, which indicates a set of characteristics in and of themselves that must be better understood.

The experience of navigating a site is expanded through the use of a predetermined path and set of points of view. Photographing becomes more even, instead of serendipitous. More ground, literally and figuratively, is covered.

Comments from my students while utilizing this method illustrate this point:

"I never would have walked over there."
"I would have avoided that part of the site."
"It looks totally different from the back."
"I noticed a new view out of the site."
"It made me stay on the site longer."[5]

The experience while photographing (as expressed by the students) becomes equally important as the photographs made. Though this method utilizes an objective structure – in that it is predetermined and may be used for any site – the process of photographing in this way is by no means objective. The process is still a unique experience for the photographer. The photographer can notice changes in their choices – *this method is making me photograph this, which I would not have otherwise. What does that mean? What can I learn about my own tendencies by engaging in this method?* Success with this method requires a combination of physical and mental preparedness. Often, moving through less than ideal conditions must occur. This physical preparation can prompt mental preparation to be attentive and open.

INFLUENCES

Bernd and Hilla Becher, founders of the Düsseldorf School, serve as a primary inspiration for this method. The Bechers began their photographic practice because of their mutual interest in documenting disappearing industrial structures throughout Germany. They started making these photographs in the late 1950s until Bernd's death in 2007. Influenced by the New Objectivity school of thought, they utilized standard, repeatable and predetermined points of view for photographing. These photographs of coal breakers, water towers and grain silos were arranged in grids, forming typologies of either one structure from multiple standardized points of view or single views of the same type of structure (i.e. water towers) grouped together.[6]

Photographed in black and white, the images were made to function like documents or records. The photographs are printed in a manner that presents the full spectrum of grays (as opposed to a high-contrast aesthetic, which favors a narrower spectrum of blacks and whites). Because of the use of an 8 x 10 camera, they are highly detailed. These visual qualities are aligned with the idea of objective documentation.[7] The regulated predetermined viewpoints do as well.[8] Often photographing parallel to the plane of the structure (like an elevation drawing) or at a 45-degree angle (like an isometric drawing[9]), they give the visual feel of a measured objective drawing.

The Bechers' photographs did not always systematically circumnavigate their subjects (they more often displayed only one or two angles). For our purposes, the approaches articulated in this chapter extrapolate the Bechers' method. Instead of collecting a set of single instances across multiple sites to create a typology (such as water towers or residential houses), circumnavigation is utilize for the creation of typology sets of a single subject of interest within a design site. For instance, all sides of a dominant feature are photographed, not just the front.[10] Similarly, circumnavigation, with predetermined viewpoints (either toward or away from the subject) at predetermined angles to the subject, regulates the decision making process for photographing. Primarily, this is about avoiding only photographing desirable views while missing other parts (such as the sides or backs) of the subject.

The effect of displaying these photographs as grids, instead of single images viewed one at a time, presents another benefit of the Bechers' practice: viewing these typologies as grids invites exploration through the process of comparison[11] (Figure 13). This comparing process stimulates looking back and forth between photographs, activating closer looking at the individual photographs. It turns into a kind of visual puzzling out. The process stimulates active looking across the set. Additionally, all sides are seen at once, or multiple views are seen simultaneously.

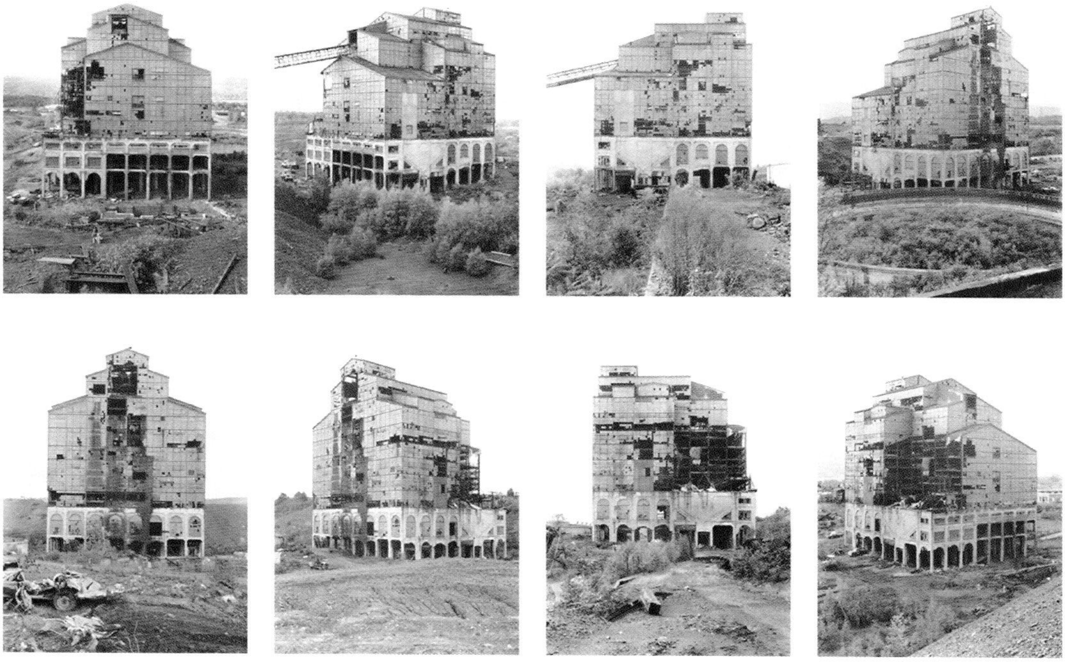

Figure 13
Bernd and Hilla Becher, Preparation Plant, Loomis Coal Breaker, Wilkes-Barre, Pennsylvania, USA 1974.

Gordon Cullen's Serial Vision, from *Townscape*, is a near neighbor to this method. He hand draws and then presents a grid of scenes depicting the navigation of urban street experiences.[12] Lawrence Halprin also utilizes a successive series of drawings depicting movement through space as part of his Scoring method, as described in *The RSVP Cycles*.[13] But both of these methods are usually one-directional. One walks through a site along a single line and looks only in one direction, in front of them. The method here unlocks the singularity of Cullen and Halprin's approach, presenting a multiplicity of views while navigating a space. This acknowledges that one looks all around, not just straightforward when moving through space. These multiple views collect in the memory, meshing with one another, which is quite different than a series of single, fixed, forward-looking scenes.

CONCLUSION

As seen in the following examples (Figures 14–21), the visual effects of this method can vary widely, thus the intentions of the photographer are always at play. As mentioned earlier, this process is both objective and subjective in nature. The use of color or black and white, the choice of focal length and the point of view are all choices the photographer makes. Yet for this method, these choices are made *before photographing the site*. That experience remains unique, even though the method is predetermined and replicable in multiple situations.

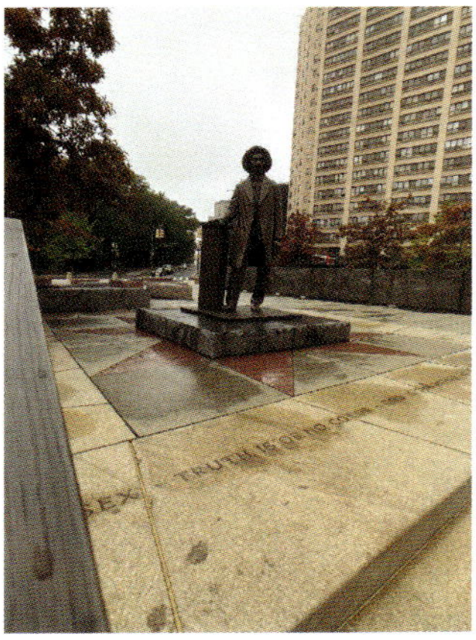

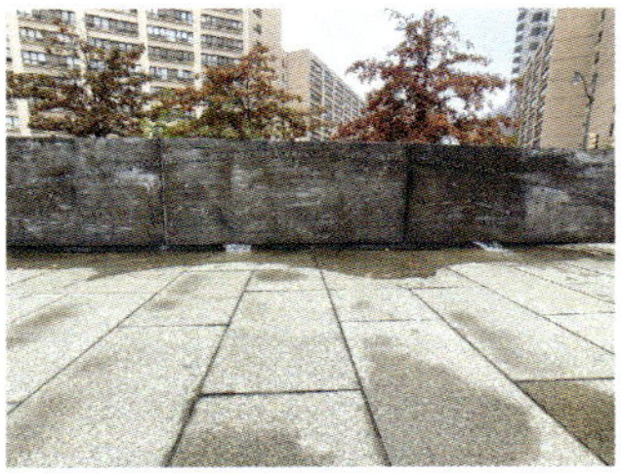

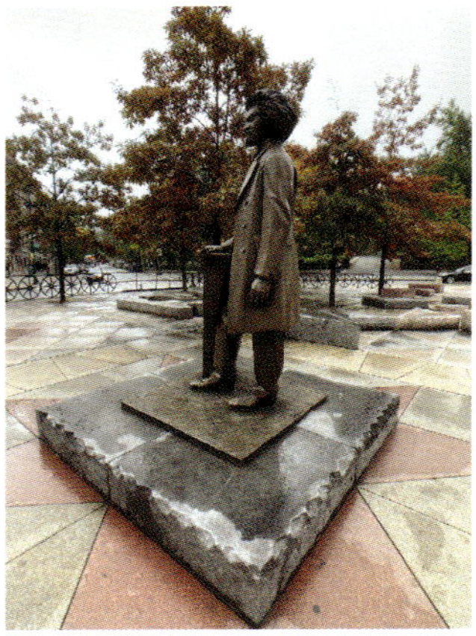

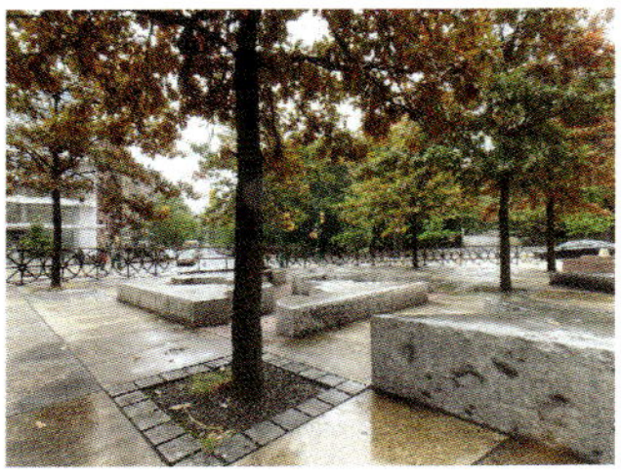

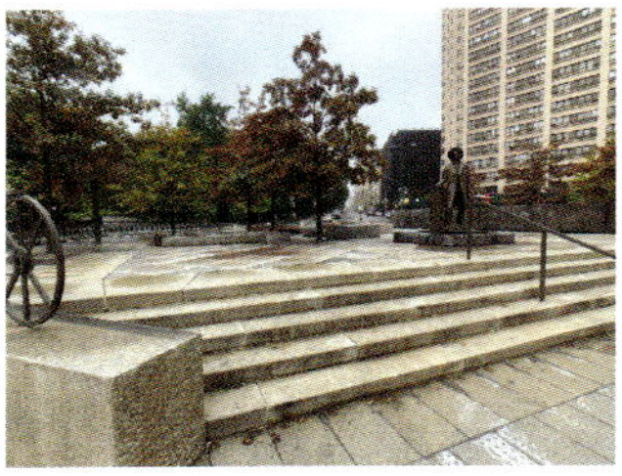

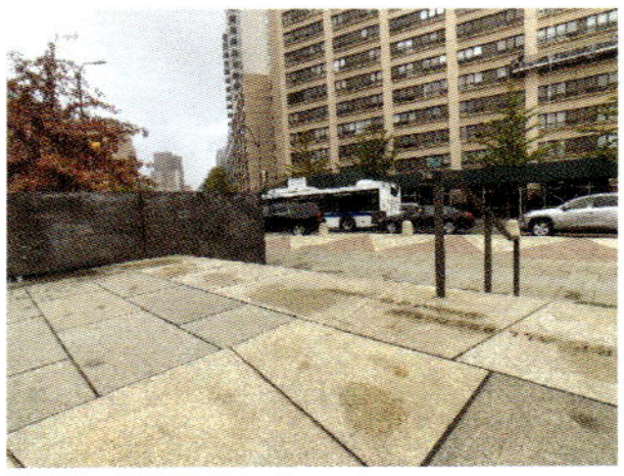

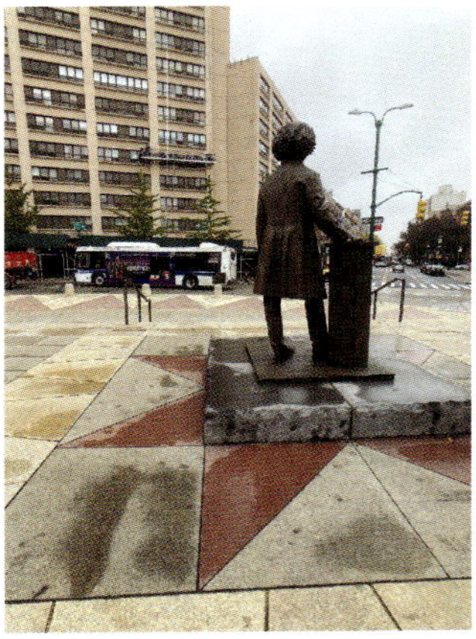

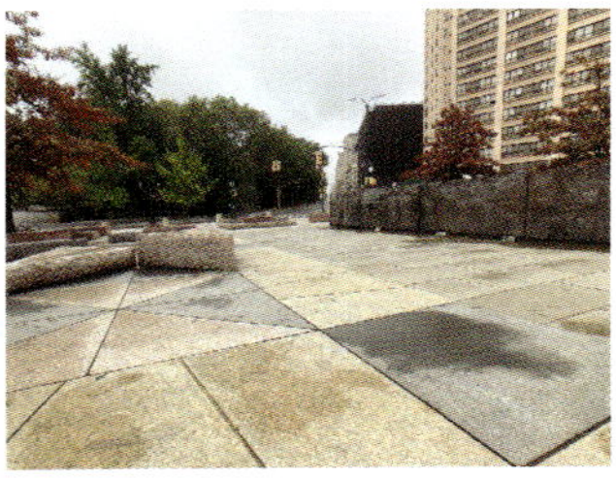

Figures 14–21
This set demonstrates a looser approach to circumnavigation. Here, at the Frederick Douglass Circle in New York, at the corner of Central Park West and 110th Street, the photographer makes one photograph toward the stature of Frederick Douglass and another looking away. Her distance from the sculpture changes for each set as she reacts to the design features throughout the circle. This approach helps the photographer explore the immediate context of the Douglass monument, instead of simply focusing on the sculpture itself as the destination.

Images: Milagros Castelan, October 2020.

NOTES

1 As well as Los Angeles and New Jersey. Though the Boston area examples are most noted.
2 Kevin Lynch, *The Image of the City* (Cambridge: MIT Press, 1960).
3 R. Gilles, "Guy Debord: The Naked City, 1957," Frac Centre-Val de Loire, last modified July 4, 2016, www.frac-centre.fr/_en/art-and-architecture-collection/debord-guy/the-naked-city-317.html?authID=53.
4 Such as 50 mm or 75 mm. These are two focal points that "simulate" how the eyes see in a landscape, but it is important to clearly acknowledge that these do not replicate actual sight in place. For more on this subject, see James F. Palmer, Vincent Vanderheyden, Gisele Alves, and Georgia Sismani, "Best Focal Length to Represent a Landscape View Using a Single-Frame Photograph," *Journal of Digital Landscape Architecture* (2017): 236–43; James F. Palmer and Robert Sullivan, "Visual Prominence as Perceived in Photographs and In-Situ." *Journal of Digital Landscape Architecture* (2017): 286–94.
5 These responses come most recently from students taking the two courses sited next, yet they have been common responses since I introduced these methods in various class and studio settings since 2005: Anne Godfrey, "Photography for Environmental Design" (Course presented at SUNY College of Environmental Science and Forestry, Syracuse, New York, Spring Semester, 2018); Anne Godfrey, "Photography, Environment and You" (Course presented at SUNY College of Environmental Science and Forestry, Syracuse, New York, Fall Semester, 2019 and 2020).
6 Susanne Lange, *Bernd and Hilla Becher: Life and Work* (Cambridge, MA: MIT Press, 2006).
7 The Bechers photographed at a time (1960s and 1970s) when the heroic photographic work of Ansel Adams, Edward Weston and their contemporaries were being reconsidered. Adams's and Weston's photography tended toward a more high-contrast palette, emphasizing deep rich blacks and bright whites. Their choice of subject matter was landscape scenes that often did not portray daily human activity in the landscape, or when it did, it was often photographed in a manner that abstracted the forms of, for example, a factory, barn, dam or human figure.

The Bechers were eventually included in a show of American photographers, *The New Topographics* (1975), who were also interested in relooking at the landscape through a cultural lens concerned with the impact of human activity on the landscape. This show exemplified a new direction for landscape photography in the late 20th century.
8 Armin Zweite, "Bernd and Hilla Becher's 'Suggestions for a Way of Seeing': Ten Key Ideas," in *Typologies*, eds. Bernd Becher and Hilla Becher (Cambridge, MA: MIT Press, 2004), 7–35.
9 This is not a true isometric since the camera lens creates recessional space, like a perspective drawing. Yet the choice of point of view, being at a 45-degree angle to the 90-degree angle of the planes of the building (when that is the case) is intentionally precise, adding to the effect of measurement and thus the effect of objectivity.
10 This also allows for seeing more of the background of a feature, as we saw in the student example preceding.
11 Anne Godfrey, "A Step Towards Multiplicity: Photography for Complexity" (Conference Presentation, Council of Educators in Landscape Architecture (CELA) Annual Conference, Sacramento, CA, March 6–9, 2019).
12 Gordon Cullen, *Townscape* (London: The Architectural Press, 1968).
13 Lawrence Halprin, *The RSVP Cycles: Creative Processes in the Human Environment* (New York: Braziller, 1981).

Practice 3 Distance

Stepping Back

DISCUSSION

The landscapes we work within are milieus of complexity. Often, they are vernacular and every day, the intermeshing of diverse uses constituting what is called the built environment. These characteristics, though considered mundane, are the very articulation of human interaction in space and place.

This matrix of infrastructural patterning usually results in a mingling of organized and haphazard physical characteristics. Again, the stuff of everyday life. To look at and examine these characteristics can require a visual assessment that reveals their essential patterns, forms, and uses. Two techniques, (1) standing back and (2) transforming details into a set of monochromatic grays, form a method for examining this complexity.

Black-and-white photography is utilized to clarify unique characteristics, while also allowing the landscape to be easily seen as a set of patterns. Instead of a high-contrast aesthetic, which emphasizes deep blacks and bright whites, a broad spectrum of grays articulates the subtleties of the landscape.

Compositionally, the tactic is to step back and show a larger portion of the landscape. This accomplishes several things. The photographic image becomes not about the representation of a single object or characteristic but instead the representation of a set of interrelated characteristics that also give a sense of the larger landscape patterns (see Figure 1). We see buildings, apartment homes, streets and parking lots. We can see the change in topography, which helps to describe the larger geomorphological context of the place. We see patterns and changes in those patterns because we step back and see more all at once. The photograph then represents a scale that describes interrelationships rather than singular details and unique small-scaled characteristics. These types of photographs are not interested in, for example, specific historic brickwork, but instead the changes in use over a greater spread of area. They function at a larger scale.

To accomplish this, the photographer gets into an aerial oblique position.[1] One takes a journey to step back and get above – a process of changing psychological position as well as physical position. Often, the photographer is no longer within, but rather outside, looking into the landscape of interest.[2]

In other cases, the photographer may not be obliquely above but is still standing in a position that is stepped back and removed. Again, compositionally, this allows the presentation of more relationships within the photograph, instead of single characteristics. The process of finding the place to step back in order to show these relationships becomes another form of assessing the landscape. *Where do I need to be to include these*

DOI: 10.4324/9780429284649-19

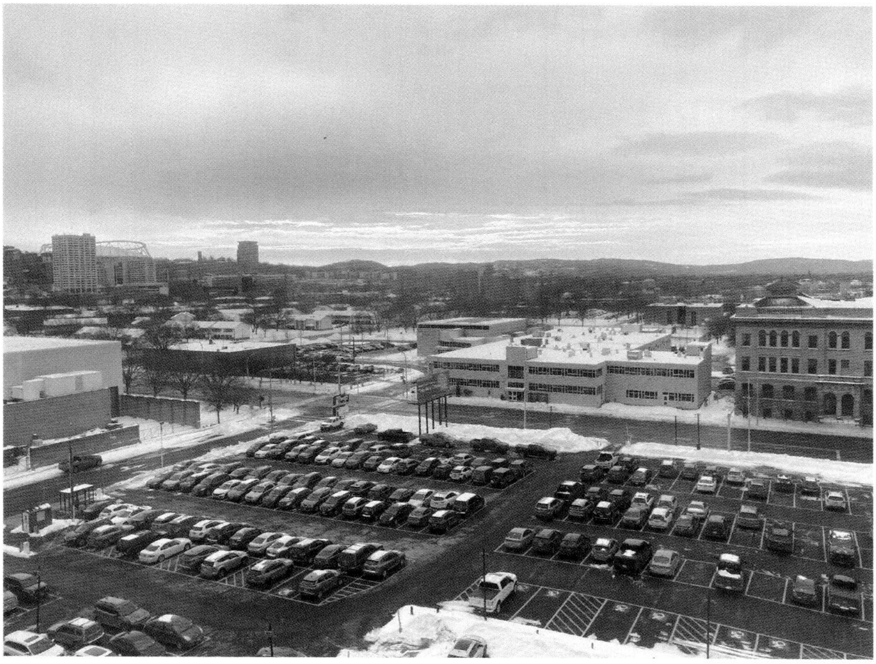

Figure 1
Downtown Syracuse, New York. This elevated view, taken from the eighth floor of the Harrison Building, looks south. Typical characteristics of this rust belt city appear in the fore, mid and background, including surface parking lots, a mix of buildings of different architectural eras and uses, the scatter of empty lots and to the left the view up to the former Carrier Dome, now replaced by a metal-fabric hybrid dome system, flanked by dormitory high rises.

Image: Anne C Godfrey, January 2022.

relationships? What do I discover through the process of finding the place that allows me to make this stepped-back view? This is another example of how the practice of a specific photographic method produces both photographs and new knowledge. The process, in this case of getting somewhere up-above or further-away, produces knowledge beyond what is depicted in the photographs themselves.

These compositional approaches produce a cool gaze – an idea associated with photographers from the New Topographics, such as Robert Adams and Joe Deal (discussed in length in the following section).[3] Observation and objectivity are central to these compositional practices.[4] In talking about his photographs of housing developments in Albuquerque, New Mexico, in the early 1970s, Deal discusses the evolution of his approach: "It was more of an accident, that I was up on a hill and looked down and could see the houses in the context of the landscape rather than just singling out the details of the architecture. I could get more distance and there was a feeling of greater objectivity in doing that."[5]

Similarly, "thinking in black and white" changes how one sees and assesses any subject matter. A photographer practices thinking in black and white while looking without photographing and imagines what the scene will look like. A visual priority is placed on form and the patterns those forms produce.

INFLUENCES

This approach to photographing is directly inspired by a group of photographers working in the 1960s and 1970s who collectively created a new genre in landscape photography: The New Topographics. Their photographs were interested in vernacular patterns in the landscape, everyday structures, and the growing sprawl of cities.

In contrast to the grand and dramatic landscape photographs of the American West made in the early 20th century by Ansel Adams, Edward Weston, Imogen Cunningham and their contemporaries, which favored majestic or sublime landscape subjects in high-contrast deep blacks and bright whites, these images presented track homes, industrial lots and cityscapes in tones favoring the mid-grays. Though people were not present in these new types of photographs, the photographs were all about people's use and activities in the everyday landscape.

The New Topographics' motivations were grounded in photographically describing from a more objective position. That position was both physical and psychological. Choosing to make compositions that included multiple features in the landscape and how they related to one another was considered a type of objective approach (see Figure 2). This is in opposition to focusing on singular heroic, highly aesthetic or highly unique features in the landscape. By stepping back, those features and the tendency to focus on those features diminishes. Instead, seen together as interrelated characteristics, the patterns become equalized within the larger landscape.

At the time, a zeitgeist was occurring around interests in the vernacular. The idealism of postwar Modernism was starting to wear, and the everyday, the real and the common were seen as legitimate subjects of interest across multiple disciplines.[6] During this time texts and works were published that explored the meaning and the significance of the vernacular, such as Robert Venturi's *Learning from Las Vegas*;[7] Ed Rusha's *26 Gasoline Stations*;[8] the *Interpretation of Ordinary Landscapes*, edited by D. W. Meining;[9] and William Whyte's *The Social Life of Small Urban Spaces*.[10]

This work was made at a time of cultural change in the United States. The promise of industrialization in postwar America was starting to crumble. Concerns arose around the pollution and degradation of land by the activities of this postwar industrial complex. Rachel Carson's *Silent Spring* was published in 1962,[11] the Love Canal disaster came to light in 1978 and suburban sprawl was transforming the western American landscape.

The pivotal New Topographics exhibit was presented in 1975, at the George Eastman House in Rochester, New York by William Jenkins, assistant curator at the time. The photographers collected and exhibited – Robert Adams, Lewis Baltz, Bernd and Hilla Becher,[12] Joe Deal, Frank Gohlke, Nicholas Nixon, John Schott, Stephen Shore[13] and Henry Wessel – each worked independently from one another in the 1960s, though some knew and were friends with one another before the exhibit (such as Deal and Nixon). Their work, individually and then collectively through this exhibit, forever changed the way we think about the depiction of landscape through photography.

As the original exhibit catalogue discussed, the photographers were not interested in the ideal portrayal of landscape as a construction made by the photographer. Instead, they strove for a more descriptive "topographic" approach to photographing the landscape.[14] The photographers both strove for and wrestled with the notion of objectivity.

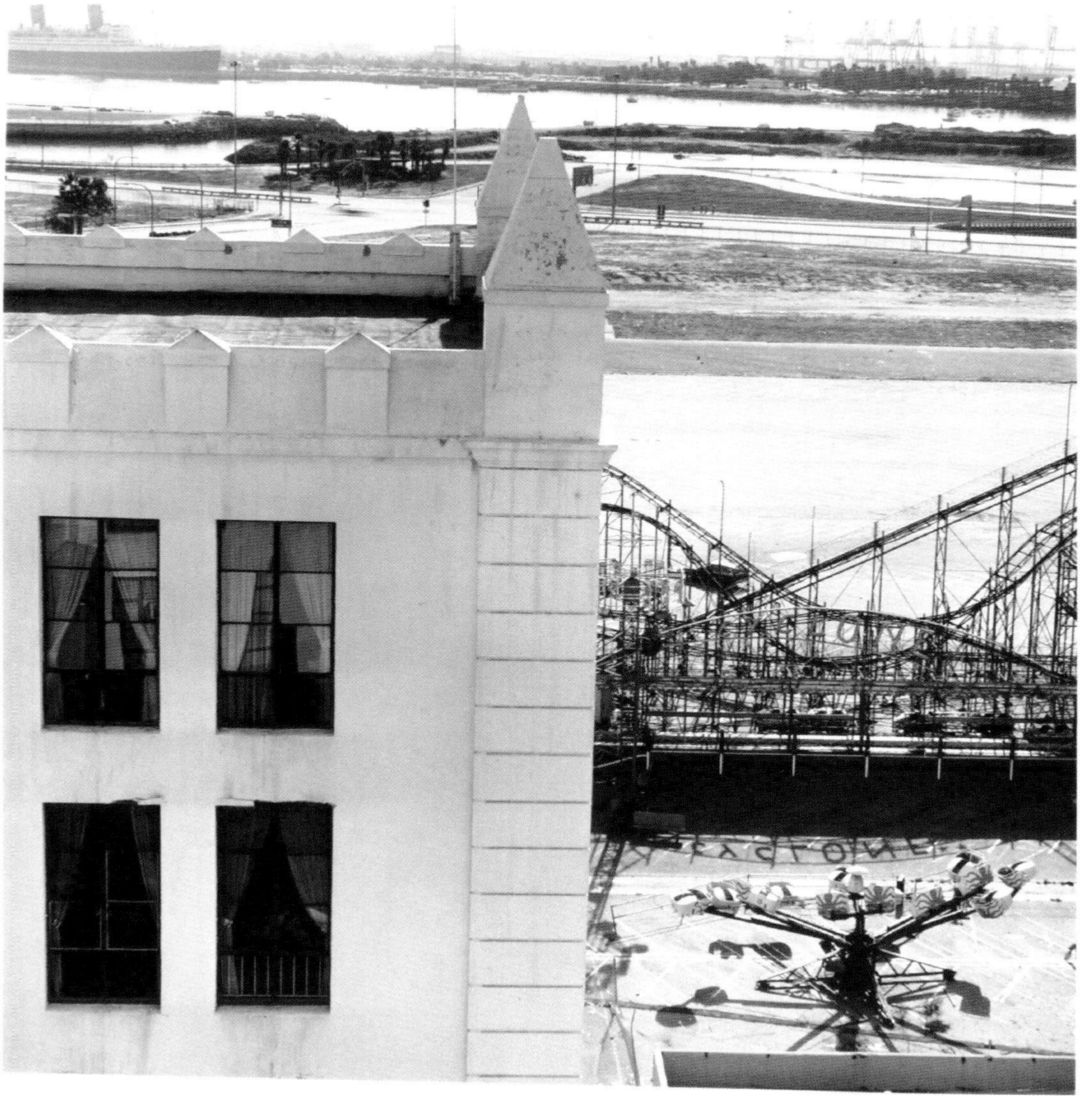

Figure 2
Joe Deal positions himself to photograph multiple uses along the Long Beach shore. The diversity of uses and structures are striking, from the residential budling in the foreground to the fair rides, shipping yard and QE2 in the distance. This composition, achieved by his elevated position, makes the image searchable.

Image: Joe Deal
Ocean Center Building and the Pike, Long Beach, California, 1980.

The concepts that formed the backbone of the photographers' working processes were simple, yet revelatory at the time. They were interested in describing. These descriptions were intricate and multilayered. The photographs presented the combined order and disorder of everyday life. They opened the door to photographically portraying these vernacular places as equal to the grand landscapes made by photographers between the 1910s and 1960s. Robert Adams stated, "That's what I'm after, a normal view of the landscape. Almost."[15] The making of these photographs acknowledged that the everyday experience of landscape was not protected National Parks showcasing unique geomorphological and ecological systems but empty lots and scrubland at the edges of expanding cities.

Today, in the early 21st century, the style of these photographs is familiar, but at the time of their presentation, they were entirely new. Prior to this show, many of the photographers struggled to have their work accepted and shown. Robert Adams recounts how his photographs submitted for review and exhibition[16] would often come back physically destroyed to demonstrate the unacceptably of his aesthetic.[17]

Joe Deal, while finishing his MFA in photography at the University of New Mexico, challenged the ideal vision of the American West by photographing the sprawl of Albuquerque and later Southern California.[18] These photographs described the reality of suburban tract homes in the American West and the attending everyday landscape experiences.

Our interests here are in the process of stepping back, observing and describing how these methods of working are related to core concepts articulated by these New Topographic photographers. Deal most clearly discusses his rationale in the original catalog of the 1975 exhibit:

> In making these photographs I attempted to make a series of images in which one image is equal in weight or appearance to another. Many of the conscious decisions made while the series was evolving had to do with denying the uniqueness of subject matter or in one exposure as opposed to another in the belief that the most extraordinary images might be the most prosaic, with a minimum of interference of personal preference and moral judgement by the photographer. An early decision was that a formal undifferentiated approach be used as a plate on which to serve up the subject matter so as to minimize the formal decisions which reoccur every time an exposure is made. The approach chosen accommodated my desire for less personal intrusion and greater uniformity. 1) By being at a greater distance from the subject matter it became difficult to significantly alter the angle of view or organization of the image by a step or two in any direction. 2) The point of view and distance from the subject allowed an acceptance by the lens of a greater amount of contextual information without allowing too great a dominance of one object over another.[19]

Deal's ongoing body of work, including his last book, *West and West* published in 2009 soon before his death, examined the details and intricate relationships to be found in both urban and rural landscapes from a removed viewpoint.[20] Often photographed either stepped back or slightly above and looking down, Deal photographed residential sprawl in Los Angeles and Albuquerque, the seismic landscapes of California,[21] the construction process at the Getty in Los Angeles[22] and his home landscape of Kansas[23] (see Figure 3). In each, the multiple layers of a place included multiple uses, multiple relationships and multiple periods of time. As he says, this is presented to the viewer in an undifferentiated way – each part of the image holds weight, instead of a strong focal point organizing the frame in a hierarchical way.

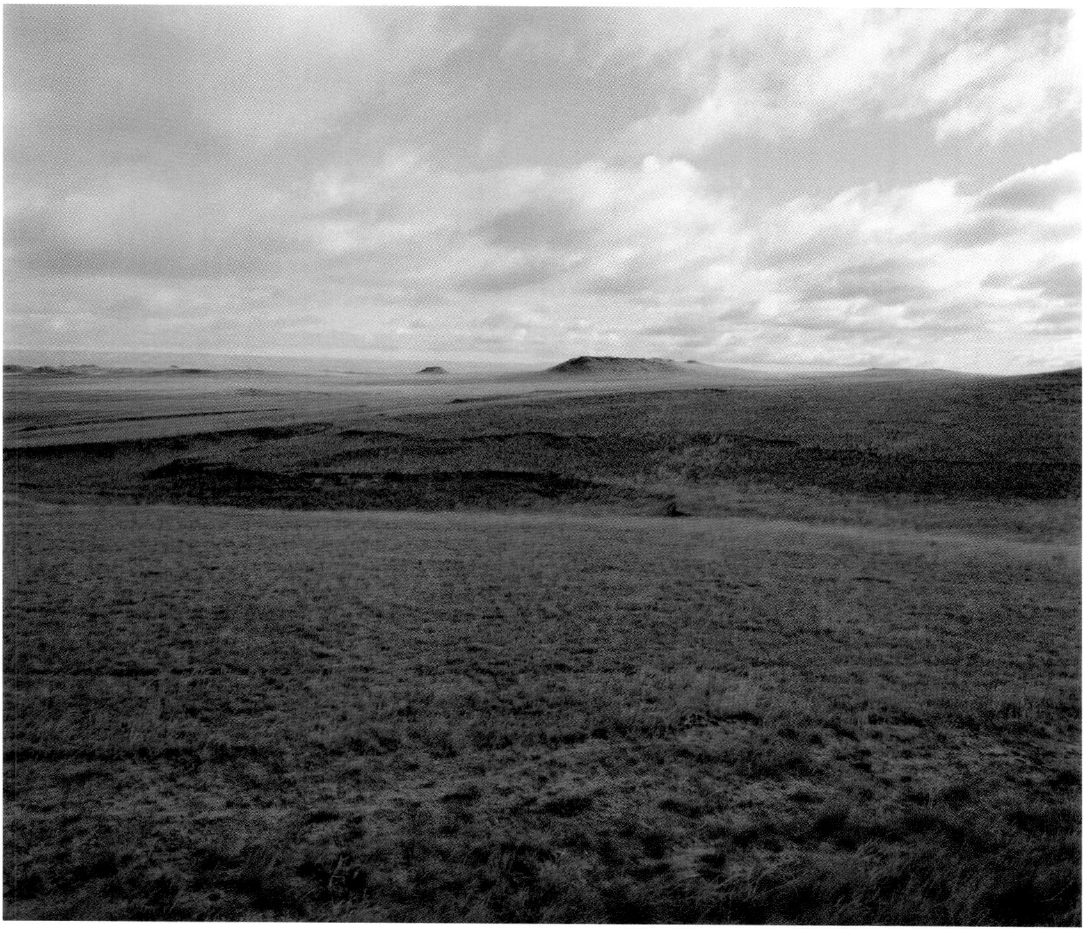

Figure 3
Deal's last published body of work took him back to his home landscape of Kansas and the expanse of the Great Plains. Though these photos do not present the same patterns found in Deal's photographs of urban landscapes, they do examine the topography of this landscape in all of its subtle changes and textures. This work is meditative as Deal reflects on the meaning of his photographic practice.

Image: Joe Deal
Missouri Plateau, 2005.
From the Series: West and West.

CONCLUSION

There is much more to the history and impact of The New Topographics than what is presented here. The New Topographics' motivations were to describe from a more objective position. That position was both physical and psychological. The objectivity comes from choosing to make compositions that include multiple features in the landscape, allowing the viewer and photographer to see how those features relate to one another. This objectivity is also achieved from standing at a distance, limiting the ability of the photographer to drastically alter the scene by small shifts in viewpoint. This is different than focusing on singular, highly aesthetic, or legible features in the landscape. By stepping back, those features, and the tendency to focus on those features, lose dominance, both visually and psychologically. Instead, they become equalized within the larger landscape because they are seen together as interrelated characteristics of a larger place. This is how concepts of objectivity are described by the photographers, especially Adams and Deal.

Multiple factors make places. By stepping back and showing those factors together in a single image – instead of broken apart, collected or made special by showing singular features or utilizing strong focal points – an evenness occurs. Things are seen in relationship to one another. This is what gets at the idea of objectivity and Deal's more specific idea of an undifferentiated approach. There is an inclusion of multiple characteristics in the frame, relaying more relationships. By stepping back and showing more, less may be left out. A greater scope of vernacular patterns is shown to articulate the functioning of everyday life.

NOTES

1 Dennis Cosgrove, "Liminal Geometry and Elemental Landscape: Construction and Representation," in *Recovering Landscape*, ed. James Corner (New York: Princeton Architectural Press, 1999), 103–20.

2 The question that comes up here is, "Why not use a drone?" It's a good question, and for some investigations, it is the right choice. Yet this method is interested in the bodily process of getting above and looking at from afar, which stimulates a psychological process. This physical process creates a different phychological experience than sending a drone up to photograph.

3 William Jenkins, *New Topographics: Photographs of a Man-Altered Landscape* (Rochester, NY: International Museum of Photography at the George Eastman House, 1975), Exhibition Catalogue.

4 Ibid.

5 Britt Salvesen, "New Topographics," in *New Topographics*, ed. Britt Salvesen (Göttingen: Steidl, 2010), 30.

6 Ibid., 35–39.

7 Robert Venturi, Denise Scott Brown and Steven Izenour, *Learning from Las Vegas* (Cambridge: The MIT Press, 1972).

8 Edward Ruscha, *Twentysix Gasoline Stations* (Alhambra: Cunningham Press, 1969).

9 Donald William Meinig, *The Interpretation of Ordinary Landscapes: Geographical Essays* (New York: Oxford University Press, 1979).

10 William H. Whyte, *The Social Life of Small Urban Spaces* (New York: Project for Public Spaces, 1980).

11 Rachel Carson. *Silent Spring* (Boston, MA: Houghton Mifflin Harcourt, 1962).

12 Discussed in the previous chapter.

13 As a contextualizing aside, Stephen Shore, the only photographer working in color, broke ground in a way by straddling the mode of street photography (which was seen as a pure expression of photography at the time, by capturing the decisive moment) and documentary description in his precise compositions of small towns across the United States. He kept one foot in Modernism's interests in the ideal aesthetic object while placing the other firmly into photographic descriptions of the mundane. Shore already had a solo show at the MoMA and had spent a considerable amount of time in Andy Warhol's Factory.

14 William Jenkins, *New Topographics: Photographs of a Man-Altered Landscape* (Rochester, NY: International Museum of Photography at the George Eastman House, 1975), Exhibition Catalogue. 5–7.

15 Salvesen, "New Topographics," 37.

16 Before the advent of digital photography, digital portfolios and digital submissions, photographers sent original photographs for review to prospective galleries and for submission to juried shows. Common practice was these portfolios were then returned to the artist. Adams became used to his work returning destroyed, cut or somehow defaced during the early part of his photography career in the 1960s. See the following note.

17 Robert Adams, "Ecology," Art 21 video, 53:36 (minutes 27:00–38:00), November 11, 2007, https://art21.org/watch/art-in-the-twenty-first-century/s4/ecology/.
18 For example, Joe Deal, J. B. Jackson, Mark Johnstone, and Edward Leffingwell, *Joe Deal: Southern California Photographs, 1976–86* (Albuquerque, NM: University of New Mexico Press, 1992).
19 Jenkins, *New Topographics*. 6–7.
20 Joe Deal, *West and West: Reimagining the Great Plains* (Chicago, IL: Center for American Places at Columbia College, 2009).
21 Colin Westerbeck, *Seismic Shift: Lewis Baltz, Joe Deal and California Landscape Photography, 1944–1984* (Riverside, CA: California Museum of Photography, 2011).
22 Joe Deal, Mark Johnstone, Richard Meier, and Weston Naef, *Between Nature and Culture: Photographs of the Getty Center* (Los Angeles, CA: The J. Paul Getty Museum, 1999).
23 Deal, *West and West.*

Practice 4 Re-Photography

More Than Comparison

DISCUSSION

Re-photography or repeat photography is a photographic practice that helps us examine change over time. The basic steps of the method include (1) finding a photograph made in the past, (2) locating the place the original photograph was made, (3) making a new repeat photograph and (4) comparing that photograph to the original photograph. Re-photography, as a common practice in many areas of the discipline, may appear simple, yet conceptually, it is quite complex.

The process of re-photography allows the designer to investigate sites and their contexts by fulfilling the predetermined goal of making a re-photograph. As with many of the methods presented here, the photographs are just one result of that goal. Again, this is a situation where the outcomes are more than the sum of the parts of the method.

Often, the focus of re-photography is on the photographs themselves. The pairs are presented to engage in a kind of A/B examination: *Let's find what is in photograph A that is not in photograph B and vice versa*. It can become a kind of curiosity, as it is fascinating to engage in the process of comparing and identifying changes between the two photographs. Implications of the changes unfurl as each person looks at the sets through their own lens of specialization. Yet this can also become an abstract activity, a cataloging of things present and absent, especially for those who have no other knowledge of the place beside the photographs themselves.

Thus, the action of comparison should be understood as just one part of this practice. The method of re-photography also directs the photographer and the design team to engage in a process of investigating the site. These investigations facilitate the locating of the specific places presented in historic photographs, and also prompt research to understand the underlying meaning (cultural, ecological, political, etc.) of the appearance or disappearance of particular characteristics. This process of investigation allows the photographer/designer to create new depth of understanding of the dynamics that have formed and changed the site, while also attaining the goal of finding the exact place where the original photograph was made. The making of the re-photographs can be treated as means to a much more complex end: an in-depth multifaceted investigation of a site, with the historic photographs and the resulting re-photographs creating a starting point, not an ending point or resolution.

Looking at the steps of this process is a way to articulate how re-photography creates new knowledge beyond the making of a new photograph:

DOI: 10.4324/9780429284649-20

Locate photographs of the site made in the past – this could include in-person visits to historic archives to finding photographs made by unidentified photographers through searching the internet.

Return to the site with those photographs in hand (most likely a reproduction suitable to take into the field). Returning can be simple, such as driving a few miles, or complex, requiring challenging travel to reach a remote site.

Find the scene presented in the photograph. No matter how simple or complex the journey, this requires some searching and navigating of the site to find not only the subject but the view and viewpoint.

Then approximate the **location of the original photographer** by engaging in on-site visual comparison. This takes the form of a feedback loop of traversing, stopping, looking at the scene, comparing it to the photograph and repeating until you find yourself in the right place. In some cases, more complex methods for locating the exact position of the camera are utilized.[1] While this is happening, the photographer is gaining new knowledge about the place through this searching process.

Make the new photograph in as close approximation as possible. This could be as simple as photographing with a phone or as complex as setting up a view camera or remote sensing instruments.

Then, later, present the two photographs together and engage in a process of **comparison** with colleagues and other interested parties. Discuss and document observed changes over time. Again, each person or group will identify, understand and interpret the photographic pairs differently based on their own background, knowledge, experience and training.

Research the conditions of change, or lack thereof, presented in the comparisons to gain a deeper understanding of what has contributed to the visual changes presented in the photograph. This research is multidisciplinary, often combining ecological history with cultural history and sociopolitical history. Though there may appear to be a single decisive narrative, most often there will be multiple histories and explanations at play. Though research appears at this point in this list, research occurs throughout the process.

Finally, the **context for comparison** will also influence the understanding of the photographic pairs. The context may be presented in a meeting, such as, "We are trying to identify what this change in topography means between these two photographs so we can make a preservation recommendation." Or the photographic pair could be presented in a community meeting: "We are interested to learn what you know about the change that has occurred here."

Over time, I have discussed this process repeatedly with Laurie Matthews, a cultural landscape preservation specialist who regularly engages in re-photography fieldwork. This work often supports re-photography practices for historic preservation endeavors interested in the protection, restoration and interpretation of cultural landscapes. As a codified method, repeat photography is used to document change over time and argue for some form of protection or rehabilitation.[2]

Matthews brings close facsimiles of historic photographs into the field, along with her camera. She creates extensive and specific field notes while she locates the original photo point. These notes go beyond recording the original location of the photographer. They include the process of finding the location, and her diverse experiences in the landscape including sounds, the condition of the soil, other people present, changes in topography and animals sited, to name a few examples. This field note-taking becomes a dynamic process of investigation that documents characteristics beyond the visual framing of a photograph. These experiences and notes assist in the discussions concerning preservation and design recommendations.[3]

INFLUENCES

Mark Klett's Rephototographic Survey Project (RSP) serves as a primary influence. The RSP team engaged in extensive field expeditions to locate and re-photograph significant landscape scenes made by the survey photographers of the American West. These survey photographers, notably Timothy O'Sullivan, William Henry Jackson, Andrew Russell and Andrew Gardener, were part of large teams supported by the United States government[4] to investigate subregions west of the 100th meridian during the late 19th century. These photographers documented unique geologic phenomena, bodies of water, land route access and other resources considered assets to be exploited by the US government.[5] These large teams included geologists, cartographers, engineers and photographers, as well as a support team of cooks, hunters and caretakers of the horses and pack animals. Each survey was named after the leader of each team – such as the King Survey, the Hayden Survey and the Wheeler Survey.[6]

The original RSP team, including Klett, Ellen Manchester, JoAnn Verburg and Rick Dingus, located and then re-made several of the best-known photographs from these surveys. The first RSP took place between 1977 and 1979, in which over 120 re-photographs were made. The primary goals of the project were to document change over time and in many instances relocate for the first time the original sites of the photographs.[7]

The RSP team utilized not only the original photographs themselves, but also maps, writings and other documents collected from the archives of the original surveys. Once the general location of the view was established, the team implemented an exacting triangulation method to locate the position of the camera and its height.[8] Through analysis of the shadows and examination of the appearance of the view in the original photograph, the team could establish the time of year and time of day when each photograph was made. They also were able to identify the original camera lenses and then used contemporary lenses as close to the originals as possible. All of this work was done with large format field cameras, similar to the cameras used by the survey photographers. But in this case, they used Polaroids and film negatives, instead of the original glass plate negatives of the survey teams.[9]

The RSP had two phases. The first set of photographs was made over three summer field seasons, starting in 1977. This resulted in the book *Second View: The Rephotographic Survey Project* (1984). The second phase occurred between 1997 and 2000, resulting in the publication of *Third Views, Second Sights: A Rephotographic Survey of the American West* (2004).[10]

The photographs are presented side by side in these books and can be interpreted from many positions, as demonstrated by the captions presented with many of the sets. They show changes in these landscapes – sometimes subtle, sometimes radical. They clearly document human alteration of the landscape, such as residential development, reservoir creation and railroad building (see Figures 1–3). These are obvious positions of

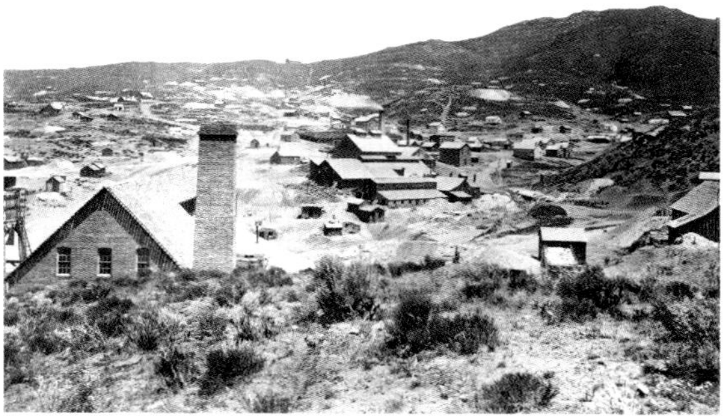

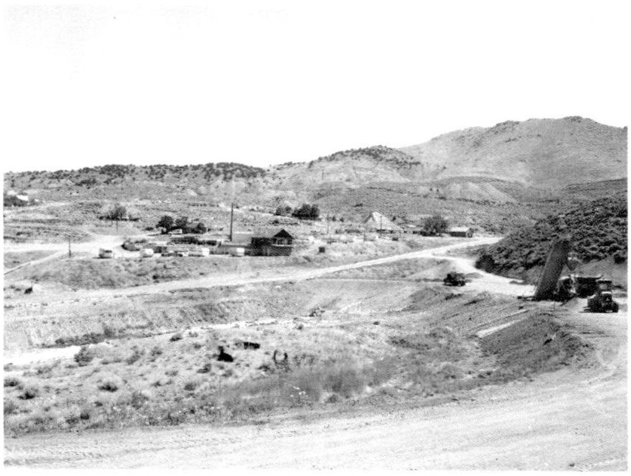

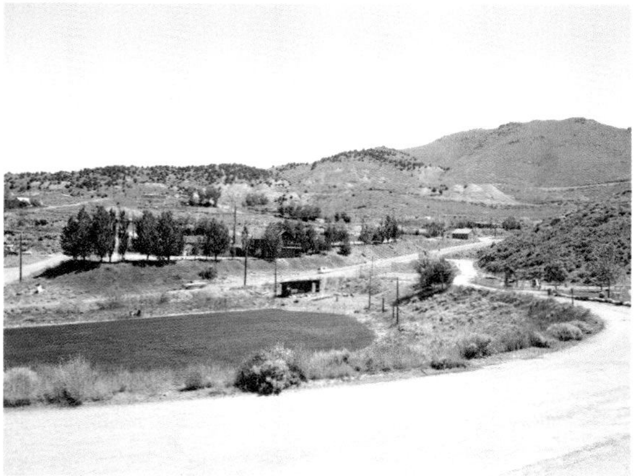

Figures 1–3
This set re-photographs Timothy O'Sullivan's 1867 photograph of Austin, Nevada (Figure 1). It was first re-photographed in 1979 by Gordon Bushaw for the RSP. The third photo, from 1998, by Mark Klett and Byron Wolfe for Third View, proved far more difficult to make. Klett writes, "The ground where the camera's vantage point stood was itself removed sometime between 1979 and 1998, when the road in the foreground was enlarged. The earlier camera positions that had been on a hillside were now literally hanging in space above the current road. To achieve the same height and vantage point, the Third View camera was set up on the roof rack of Kyle Bajakian's truck camper shell and literally driven into the correct position."

Images: Timothy O'Sullivan, 1867 (U.S. Geological Survey); Gordan Bushaw, 1979; Mark Klett and Byron Wolfe, 1998. Figures 2 and 3 courtesy of Mark Klett.

interpretation. By assessing these changes, questions arise. *What are the implications of the changes apparent in these photographic sets?*

Less obvious positions for interpretation lie in the unpredictable discoveries and experiences that occurred while making these re-photographs. Both RSP books present the sets of photographs with narratives describing experiences while working in the field. A diversity of unique and unplanned encounters are shared, including daily challenges of making the photographs, contemporary insights into the work of the survey photographers and inclusion of personal work made by each of the team of photographers. For the second RSP (Third View), Klett created a companion interactive website, including posts with updates from the field, map points displaying the clustering of the sites and antidotal experiences. While the Third View project was underway, writer William L. Fox joined the team and created his own companion book, interweaving stories from life in the field with a discussion of the history of photography in the American West.[11]

From this position of interpretation, we see how re-photography as a method includes contemporary experiences and processes in the making of the photographs that also inform the understanding of these landscapes. This method becomes about much more than understanding the choices and influences of the original photographers. When this type of external information is included, re-photography becomes far more than visual A/B comparisons.

For example, RSP experiences in the field led to a new understanding of the real scale and meaning of certain now famous original survey photographs of geologic formations (such as Witches Rock and Camel Rock). The original survey photographers often photographed extraordinary geomorphological phenomena. These phenomena were considered unusual and of great interest to the geologists on the teams (especially the King Survey[12]). Framed and composed in ways that set them into what appears to be empty expanses of the arid western landscape, these features look huge. But the second and third views of the RSP reveal (through the appearance of other human-made objects) how the understanding or perception of the size of these features was sometimes exaggerated. The framing and camera position in the original photographs exclude contextual features that may have better scaled the formations. Thus, these original survey photographs cannot be understood as static, unauthored visual copies of these places. Klett explains,

> When we look at 19th century geologic survey photographs, we are often unaware of individual preferences exercised by photographers and are tempted to assume that these images are faithful documents of rock strata in the Wild West. By finding where these photographs were made we were often surprised by the highly individualized decisions made by early photographers.[13]

These multiple experiential and narrative contexts, shared in essays and extended captions throughout the two books, demonstrate how re-photography can become a process of discovery that extends beyond viewing the resulting photographs. Verburg writes,

> Comparing the original survey photographer's view to the site where it was made was by far the most provocative aspect of our field work, first, because it was always a shock

> to see that the place didn't look like the picture, second, because we could compare what was included in the frame with what was left out, and third, because it was a way to experience 100 years through a very simple comparison.[14]

Looking at the Shoshone Fall's series, for example, serves as an example of how these contexts may be shared, explaining the experiential and narrative, as well as the visual (see caption for Figures 4–5).

Though the strict method of locating and photographing strives for a kind of objectivity, because of the commitment to precisely locating the original camera position and photographing at the same time of day and year, the daily life in the field is steeped in unique human experiences and emotions. This daily interaction, in sharp focus during fieldwork, sits side by side with the methodological process of the re-photography. Klett reflects on how the fieldwork for the RSP solidified his own thoughts on landscape photography. In *Third View*, he presents three observations:

> First, landscape photographs result from an interaction of personal experience and cultural influences. They reflect the external forces that shaped the vision of their makers, and in similar ways, the forces that shaped viewers influence their interpretation of photographs. Second, photographers are always participants in the making of a landscape, not simply impartial witnesses, as their photographs are never neutral. Third, landscape photographs are timely even when they seem timeless, and the moment of each image is unique and can never be exactly repeated.[15]

CONCLUSION

The methodologic practice of re-photography can assist in understanding multiple layers of landscape history and experience. Re-photography allows for visual comparison of change over time in any landscape. The process of re-photographing directs the designer/photographer through a site in specific ways, as the goal is to locate the "original photograph." This process is also about what is to be learned while searching for those original photo points. Many types of valuable experiences occur throughout the process of successfully making the re-photograph.

The examples presented here aim to understand change over the course of decades or even a century. Yet re-photography can be utilized for understanding change over much shorter amounts of time. For example, re-photography could be used to look at how a gentrifying neighborhood has changed over three years. This kind of re-photography could include notes about daily interactions on the sidewalk, how people decorate their yards or how many children are out playing on a Saturday. Another example could look at catastrophic storm recovery, starting with photographs made after the destructive storm and then future photographs documenting recovery (or lack thereof) or implementation of more resilient Wolfe shore edge interventions. The notes and process of making these photographs could center on water flows, species counting, humans' specific activities or where more or less recovery and cleanup is occurring and why.

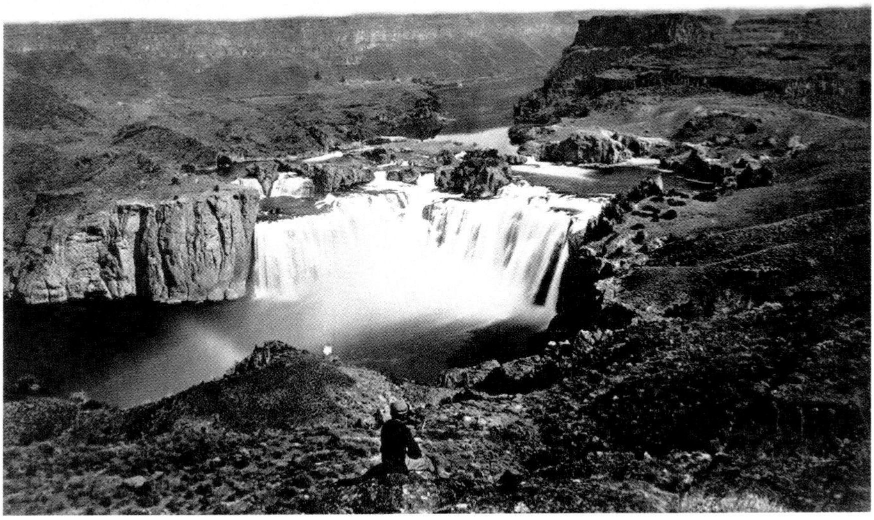

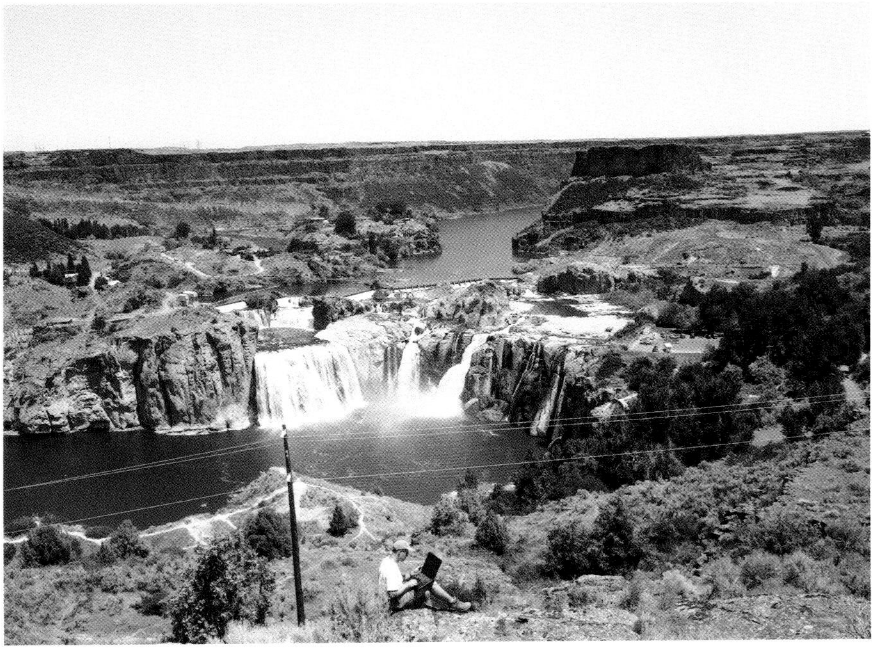

Figures 4–5
Timothy O'Sullivan's 1868 photograph of Shoshone Falls in Idaho was re-photographed in 1999 by Mark Klett and Byron Wolfe for Third View. If a human figure appeared in the original photograph, the team often placed a contemporary human figure into the re-photograph. The man in the first photo is either drawing or writing, while the contemporary man is doing a contemporary version of that activity, working on a laptop. In this case, the man is viewing a digital version of the original photograph downloaded from the National Archives.

Images: Timothy O'Sullivan, King Survey, National Archives; Mark Klett and Byron Wolfe, 1999. Re-photograph courtesy of Mark Klett.

In each of these cases, the re-photography method produces both photographs and on-site experience during the process of making the photographs. Combing these two sets of knowledge creates a more complete story about how and why places change over time.

NOTES

1 Mark Klett, "Three Methods of Presenting Repeat Photographs," in *Repeat Photography: Methods and Applications in the Natural Sciences*, eds. Robert H. Webb, Diane E. Boyer, and Raymond M. Turner (Washington, DC: Island Press, 2010), 32–45. Klett's chapter serves as one example. This entire volume articulates a set of specific methods for location of station points in different settings and for different uses.
2 National Park Service, "Heritage Documentation Programs: HABS, HAER, HALS," October 21, 2021, www.nps.gov/hdp/.
3 One significant discussion cited here: Roundtable discussion. MIG. Portland, OR, August 2018. Including: Laurie Matthews, Casey Howard.
4 Two of these survey teams were part of the precursor of the current US Geological Survey – better known as the USGS.
5 One discussion of the motivations of the surveys is found in Martha A. Sandweiss, *Print the Legend: Photography and the American West* (New Haven, CT: Yale University Press, 2002).
6 Respectively, Clarence King, Ferdinand Vandeveer Hayden, Lieutenant George Montague Wheeler. Specifically, the King and Hayden Surveys were conducted under the USGS, while the Wheeler Survey was undertaken by the Army Corps of Engineers, though the three are often considered together as the significant foundational government surveys of the American West. It should be noted that these surveys, in part, assisted the US government to continue to appropriate land from the indigenous peoples of these areas.
7 Mark Klett, Ellen Manchester, JoAnn Verburg, Gordon Bushaw, Rich Dingus, and Paul Berger, *Second View: The Rephotographic Survey Project* (Albuquerque, NM: The University of New Mexico Press, 1984).
8 Mark Klett, "Rephotographing Nineteenth-Century Landscapes," in *Second View: The Rephotographic Survey Project*, eds. Mark Klett, Ellen Manchester, JoAnn Verburg, Gordon Bushaw, Rich Dingus, and Paul Berger (Albuquerque, NM: The University of New Mexico Press, 1984), 11–44.
9 Klett likens drying washed polaroid film, hung on clothesline in the back of his van, to the drying process of developing and fixing glass plates. Ibid.
10 Klett et al., *Second View*; Mark Klett, Kyle Bajakian, William L. Fox, Michael Marshall, Toshi Ueshina, and Byron Wolfe, *Third Views, Second Sights: A Rephotographic Survey of the American West* (Santa Fe, CA: Museum of New Mexico Press in association with the Center for American Places, 2004).
11 William L. Fox, *View Finder: Mark Klett, Photography, and the Reinvention of Landscape* (Albuquerque, NM: University of New Mexico Press, 2001).
12 The King Survey, which surveyed the 40th parallel west of the 100th meridian of the American West, was led by Clarence King, Geologist in Charge of the survey of the 40th parallel. Trained as a geologist at Yale, he eventually became Director of the USGS.
13 Klett, "Rephotographing Nineteenth-Century Landscapes," 17.
14 JoAnn Verburg, "Between Exposures," Essay, in *Second View: The Rephotographic Survey Project* (Albuquerque, NM: The University of New Mexico Press, 1984), 5–10.
15 Klett et al., "Introduction," *Third Views, Second Sights*, 3.

Practice 5 Community

DISCUSSION

There are many ways to engage community. Several methods from the social sciences, such as the multiple methods within the practice of Photovoice,[1] include photography as part of outreach efforts. These methods represent one set of approaches utilizing photography as part of community outreach.

Alternatively, the methods in this collection are interested in the practices of photographers in the arts. Here, practices in portraiture guide this method. What lessons can landscape architects and environmental designers learn from different contemproary approaches to portraiture?

First, photographing people is one way to understand place. People express the landscape. People's gazes, expressions, gestures and postures are influenced in part by their engagement with the landscape. People translate the landscape through stance, clothing and the wear and tear of their bodies. People refract the landscape.

Dorothea Lange's photographic portraits made during the Dust Bowl for the US government's Works Progress Administration tell the story of individual human struggles and the devastation of the landscape through nonregenerative agricultural practices.[2] Naima Green's contemporary portraits of Black subjects in outdoor green spaces flip expectations for where particular human subjects are expected to be found and enjoy spending time.[3] Though these two examples are quite different from one another in motivation and time period, human and landscape subjects tell complex stories together.

Common practices in architecture and landscape architecture use people as props to demonstrate use. The people are doing things that show off the success of the design, or show, in the case of existing conditions photographs, what may or may not be working. Yet do we see these people as individuals when we photograph them? Or do we clump them into groups, demographics, or simply aesthetically pleasing features in our photographs? Practicing portraiture is a way to break through some of these unexamined habits.

At our best, we cultivate meaningful interaction between ourselves and the people in these places. The actions of photography can become more than transactional. I talk with and get to know that person. I cultivate empathy with the people who populate my photographs through interviews, research and casual conversation. Why are they here? What are they doing? How are they feeling? What do they want? What do they think about this place?

DOI: 10.4324/9780429284649-21

As I have written in *Landscape Architecture Magazine*,[4] the places we make for people, especially public spaces, become places that enliven specific interpersonal moments. A wonderful place to sit in the sun on an early spring day, especially after a long dreary winter, becomes a place to have a specific conversation. A path to stroll easily and safely may become a turning point for someone's health. These design goals might be general, yet they have unique outcomes for specific people. We make places, but what really happens is that we make opportunities for the experience of personal moments. Our work serves as both facilitator and backdrop to the everyday process of life. Sometimes those events are mundane and repeated, while others are unique and life changing.

A way to start seeing the usefulness of portraiture is to equate altering the landscape through design with altering people's everyday lives. When we make places in any landscape setting, our work affects people's daily activities in small and large ways. These alterations can create opportunities for interaction, shift locales of gathering, change air or water quality or transform a desolate place to a welcoming one.

How can practices with portraiture help us gain a better understanding of the people whose lives will be affected by our work?

INFLUENCES

Since the beginning of his career, Dawoud Bey has photographed people. From street photography portraits, such as his early work in Harlem in the 1970s, to more intentionally orchestrated portrait sittings with high school students in the body of work *Class Pictures*, Bey's art examines the relationship between photographic intention and description. He treats his relationship with subjects as reciprocal, not one-sided. He considers the photographic process of portrait making as something that alters each of their lives – the photographer's and the subject's.[5]

Bey makes photographs of people in many types of communities. He is closely tied to some of those communities, while for others he has had little previous experience and comes to them as an outsider. Considerations are to be made in either situation:

> No matter your position, inside or outside, it's a question of intention. There also has to be critical thought applied to that intention. One needs to critique one's own motivation. Not all intentions can be justified. I'm not saying don't make any photographs, but there has to be a moment when you step back and ask what this photograph means outside of what makes for an "interesting picture."[6]

Landscape architects face similar situations. Sometimes we design within a community we are closely related to, while other times we work in a place where at the start we have little knowledge. While working in any of those settings, we make photographs of people in place. Bey presents a position from which to assess our photographic intentions. Are we photographing people because they illustrate an interesting point or make for a good picture? Or are we utilizing photography to gain some kind of deeper understanding of these unique people's needs? By simply raising awareness about our intentions, the type of considerations Bey suggests become available. Do I want this person to be in the photograph because of their colorful shirt? Or do I want this person in the photograph because they are doing a particular unique activity that is important to my understanding of how this place is used?

Class Pictures, a project with its roots in a residency at the Addison Gallery at Phillips Academy, placed Bey in a context where he was an outsider. Phillips Academy, a prestigious and wealthy private boarding school in a small town in Massachusetts, differs greatly from Bey's adopted home of Chicago and his academic institution, Columbia College. Yet the project was not predicated on the outsider looking in but rather on how as a photographer Bey could find and express complexity and uniqueness in his subjects, teenagers, whom he believes is a group of people often abstracted and kept two dimensional. This work with the Addison Gallery expanded into a multiyear project, taking Bey to multiple educational settings to photograph teenagers from many walks of life. Bey discusses,

> I continued to be interested in teenage subjects because I believe that this society, in spite of all its rhetoric, really doesn't care about young people as much as it purports to. The culture tends to come at young people in ways that are useful for an agenda, representing teenagers as either socially problematic or considering them as engines for consumerism. They have been generalized and stereotyped in just as many, though different, ways as Black subjects – neither are depicted as having rich inner lives. I wanted to construct a representation of teenagers in opposition to that, coming to them as a way of finding out what they're really thinking about. And I wanted to make portraits across a broad range of educational cultures to create a more complex visualization of young people than the rather one-dimensional portrayal we see in popular culture.[7]

Bey hits upon an important issue, how representations may or may not allow for the interior life of the subject. Bey strives to reveal that interiority. For *Class Pictures*, Bey engaged with the teenagers in multiple ways. In some instances, the students are interviewed (sometimes by Dan Collison), and in other settings, Bey has asked the students to write about a question that asks them to describe "one's private self that is not necessarily on public display."[8]

Later, the subject's written, recorded or transcribed words (depending on the setting) are adjoined with Bey's portraits. The photographs and text combine to disrupt the stereotype of teenagers and put into question the biased eye of the viewer. Bey says, "I want to try to make the fullest, most interesting and compelling visual description of the person and then, hopefully, when you put the two of them together – my visual voice with their literary voice – you end up with this more dimensional third thing."[9]

Our work with diverse communities, both familiar and new, can be informed by Bey's approach and intentions with the teenagers for *Class Pictures*. As a way to wrap our arms around the initial issues of a project, we often create labels or containers for groups of people in communities. These labels, though not intended to be disparaging, do abstract individual experience into general concepts: working class, immigrant community, white middle-class suburban, elderly, grade school children, tourists, patrons, government officials, displaced by gentrification, developers, landlords, underemployed, rich, service industry workers. Labeling gives us a place to start, like teenagers. But it also marks a place to end, too quickly. The practice of portrait making creates an opportunity to transcend simple labels.

Making a portrait requires the photographer to get close to the subject. A portrait is defined by focusing on the body of the subject, most often the head and torso. The body fills a good portion of the frame. And for Bey, the primary motivation of the portrait is to describe in a way that clearly understands the underlying intention.

Cultivating an understanding of intention can actively direct photography for design. How do we treat human subjects as we make photographs of them as part of the design process? Bey's own thoughts on this introduce a question: *what are my intentions?* Asking one of the core questions from the first section of this book, *why am I making this photograph?*,[10] also activates the photographer/designer to examine their intentions:

> Am I making these photographs of people because I find them to be aesthetically pleasing?
> What are the implications of that intention?
> Or perhaps, does making these photographs help me understand a community event or value?
> How might a clarified statement of intention change my approach to photographing?
> How can I photograph in a way that cultivates a sense of empathy and understanding of the people who call this place home?

With clearly articulated intentions, made and discovered through the process of photographing, the abstracted labels we grapple with can transform into unique stories and experiences. The process of making portraits is one of meeting, talking with and spending time with the subject. For Bey, this is a process of revealing that helps him come to a better understanding of the person. With *Class Pictures*, that understanding is passed on to the viewer through the inclusion of the subject's spoken or written words about themselves. Bey succeeds in his intentions – acknowledging the complex inner life and self-reflection his subjects possess.

Painter Amy Sherald also examines the idea of intention and the ability of photography to reveal and acknowledge inner life and personal agency. She discusses how personal photography, or snapshot photography, allows for the representation of interiority in Black subjects.[11] Instead of the gaze of representation being directed by an outsider or a person holding power (in more formal settings such as painting, film or commercial photography), the gaze in personal photography is created by an intimate everyday two-way relationship: a snapshot made of a son by a mother, a girlfriend by a lover, a cousin by another cousin. In these casual settings, both the subject and the photographer can express their own personal and interior spaces.[12] The body is relaxed, the stance casual and the gaze into the camera is open, trusting, unguarded. Thus, Sherald sees personal photography as the form of representation that allows for the expression of Black interiority and complexity. She states, "It gave people who once had no control over the proliferation of their own image the ability to become authors of their narratives."[13] Sherald discusses how her paintings are deeply inspired by this personal photography. Her own portrait painting relies on this type of photography for inspiration, as she collects and makes photographs made in this mode as part of her artist's practice.[14]

Naima Green, introduced earlier, explores how portraiture can disrupt common and unexamined stereotypes. Her ongoing body of work, *Jewels From the Hinterland*, presents Black and Brown bodies in green spaces. Green talks with people from her home

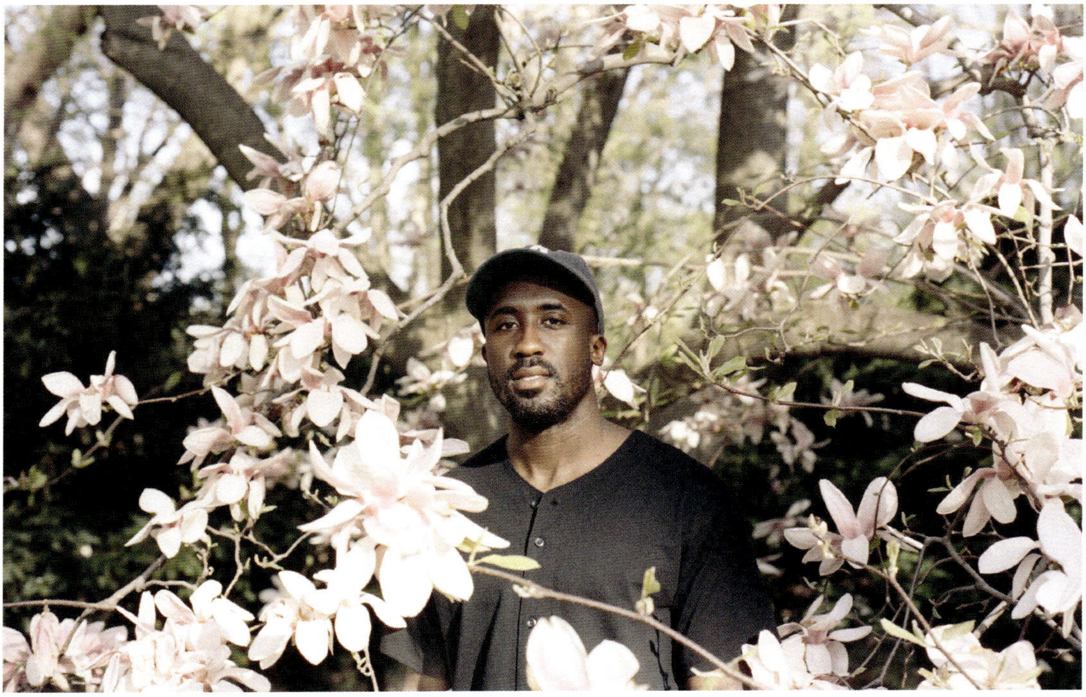

Figure 1
Jewels From the Hinterland, Lipton, Prospect Park, 2015.

Image: Naima Green.

community or people she meets while traveling and asks them to take her to their favorite outdoor places – urban parks and green spaces found across the United States (see Figure 1). The subjects stand among tree limbs, relax in tall grasses or lean against rocks along a shoreline. This displaces the often unsaid expectation of seeing Black people surrounded by concrete and urban infrastructure. Green explains, "So many of the portraits are about confronting the viewers' ideas that Black people can't exist within lush green urban environments as spaces of leisure and rest."[15] She wanted to reclaim "lively lush urban space as Black space," with this work.[16] Though Green's portraits do not include narratives like Bey's *Class Pictures*, they do support the expression of unique interiority that Sherald believes is available through the process of intimate portraiture photography.

CONCLUSION

The uniqueness of human experience often motivates our work. What if we made more time and space to represent those specific stories? Portraiture can support an intentional photographic process that uncovers complexity, uniqueness and interiority. It can also help us, as demonstrated by the previous examples, remove the stereotypes we create when we put people into categories.

The following book in this series *Active Landscape Photography: Diverse Practices*, presents a chapter written by photographer Sahar Coston-Hardy and urban planner

Rhiannon Sinclair. They discuss their community engagement process for Agency Planning + Design's XX Miller Prize design intervention at the AT&T Building in Columbus, Indiana. Coston-Hardy and Sinclair recorded interviews and made thoughtful portraits, creating a powerful process for engaging the community. This example illustrates the common interrelationship between a landscape place, its multiple histories and the unique stories of the people who care about it.

NOTES

1 For example, see Amanda O. Latz, *Photovoice Research in Education and Beyond: A Practical Guide from Theory to Exhibition* (New York: Routledge, Taylor & Francis Group, 2017).
2 For example, see Anne Whiston Spirn and Dorothea Lange, *Daring to Look: Dorothea Lange's Photographs and Reports from the Field* (Chicago: The University of Chicago Press, 2010); Dorothea Lange, *Dorothea Lange: Words & Pictures*, New York, Museum of Modern Art, February 9–September 19, 2020, www.moma.org/calendar/exhibitions/5079.
3 Shikeith and Naima Green, "Aperture Conversations: Antwaun Sargent, Shikeith, and Naima Green on Black Queer Utopia," interview by Antwaun Sargent, Aperture Foundation, February 19, 2021, www.youtube.com/watch?v=M8PiD2xRkul.
4 Anne Godfrey, "Tanner Springs Moments," *Landscape Architecture Magazine*, February 2020, 110–16.
5 Dawoud Bey and Brian Ulrich, *Dawoud Bey on Photographing People and Communities* (New York: Aperture, 2019).
6 Ibid., 23.
7 Dawoud Bey, *Dawoud Bey: Seeing Deeply* (Austin, TX: University of Texas Press, 2018), 85.
8 Dawoud Bey, interview by Rebecca Dimling Cochran, "Interview: Photographer Dawoud Bey on 'Class Pictures' and the Emory Project," *ArtsATL*, July 20, 2010, www.artsatl.org/a-conversation-with-photographer-dawoud-bey-about-class-pictures-and-the-emory-project/.
9 Ibid.
10 See Method 3 Asking questions.
11 Amy Sherald, *The Heart of the Matter* (New York: Hauser and Wirth Gallery, September 10, 2019–October 26, 2019).
12 This is to say the type of personal photography Sherald is talking about is different than a selfie or other photographs made explicitly to be shared on social media.
13 Ibid., see website: "Amy Sherald the heart of the matter," Hauser and Wirth, accessed November 2021, www.hauserwirth.com/hauser-wirth-exhibitions/24653-amy-sherald.
14 Amy Sherald and Kehinde Wiley, interview by Doreen St. Felix, "Brooklyn Talks: Amy Sherald and Kehinde Wiley," Brooklyn Museum, October 14, 2021, www.brooklynmuseum.org/calendar/event/brooklyn_talks_sherald_wiley_october_2021/.
15 Shikeith and Green, "Aperture Conversations."
16 Antwaun Sargent, "The Future Will See You Now," *Aperture 241 Utopia*, December 8, 2020, 91.

Practice 6 Things Take Time: Time Takes Things

DISCUSSION

Single photographs display the success of our work. Yet underneath those "final" images exist the long-term relationships designers and researchers develop with sites and their larger contexts. Projects can last for years, resulting in multiple visits, multiple photographs and multiple experiences. We become attached, as well as keenly aware of the specific details and circumstances of the place. Our relationship with these places often extends beyond the goals of the project. We may continue to visit again and again, long after the project is finished. This act of revisiting occurs with other places as well – places we did not design but have studied, used as precedents, admire or find personally significant.

As time passes, our attitude or relationship with a particular place shifts. Our interest in photographing that place also shifts. This method utilizes a looser form of re-photography, inspired by William Christenberry, which allows for idiosyncratic positions (point of view) for the repeat photography. This approach reveals both changes in the site over time and the changing experiences and values of the photographer(s). Assembled as a whole, these photographs become a narrative as well as a visual analysis.

Processes of change express themselves through physical (and thus visual) manifestations. The photographer is motivated by these physical occurrences. Also, the photographer brings their own set of personal changes, accumulated over time, to the process of photographing. A commitment to a long-term photography relationship with a landscape place can allow for the idiosyncratic interplay of these two conditions.

A dialogue transpires between the photographer/designer and a place over time. How I see and understand a landscape place today is different than how I will understand it ten years from now. The way I photograph that place often reflects that change. What I frame, the point of view I utilize and the focal point I create will likely shift to reflect those changes. The time of day and the time of year the successive photographs are made express when I am able to visit the place (which might be determined by other contextual factors, such as visiting friends or family, vacation time or simply when the project cycle marks another trip to the site). Looking at these photographs collectively, later, can become an introspective process. This is decidedly different than the re-photography approach discussed in Practice 4. This method allows the photographer to photograph the same place over time in different ways, from different viewpoints, with different compositions and at different times of day and year.

This method also allows for the expression of time taking things. Characteristics of a place grow, change, break, leave, disappear, accumulate again, are renewed,

DOI: 10.4324/9780429284649-22

replaced, repositioned and then deteriorate, get worn and disappear again. The sites we design take time to mature, grow, become more rigorously visited, loved, valued and then wear, need repair and change again. Time takes and transforms. This relates to the primary goal of revisiting and re-photographing: to see and document change. Yet change over time is not linear, nor does it trace the same trajectory over and over. Change can be unpredictable. Time takes, replaces and alters.

INFLUENCE

Since 1961, photographer, painter and sculptor William Christenberry (1936–2016) had returned to his home of Hale County, Alabama, to photograph and re-photograph places and structures that epitomized, for him, the vernacular landscape of the American Deep South. His interests were in the everyday places that change and disappear or change but remain seemingly the same over time. His longitudinal photographic studies of these places span not only years but also decades.[1]

This work was motivated by an ongoing personal dialogue Christenberry maintained with his home landscape. "I miss living in the Deep South a lot. I have such an attachment to it and such a deep feeling about it, but it's how you respond to it – how you respond to what you are about. I like the sense of distance and mystery that generates feelings by living outside of the Deep South."[2] This relationship is not about unexamined nostalgia for this place, but rather his ongoing complex dialogue with a cultural landscape that he both missed and critiqued.[3]

Christenberry assembled these collections of photographs into grids or side-by-sides so that both he and the viewer could see and experience the changes over time. The primary subject matter, often a building, appears differently in each photograph. In the *Green Warehouse, Newbern, Alabama*, 1973–2004 (21 ektacolor prints), Christenberry alternatively fills the frame with the green warehouse or steps back, showing a larger portion of the landscape. Sometimes he stands parallel to the face of the building, while other times, he places his camera's plane slightly askew. The collection is as much an expression of Christenberry's feelings and experiences in the moment of photographing as they are a document of change over time for the structure of interest. *The Red Building in Forest, Hale County, Alabama*, 1974–2004 (20 pigment prints), is similarly photographed, showing some intermittent changes, yet staying very much the same over time.

At other times, these buildings changed radically, such as *The Underground Club, Greensboro Alabama* 1967–2005 (21 ektacolor prints, printed 2006). For this series Christenberry's framing of the building remains virtually the same, standing in front of and parallel to the building, almost like an elevation view. The building goes through multiple names, is torn down and rebuilt and turns from red to blue.

Christenberry describes why he maintains this ongoing relationship with these structures:

> I can't say I'm obsessively attracted, but honestly I find beauty in things that are old and changing, like we all are changing. I don't find them ugly. I certainly don't think I'm making a picture of a "decaying South" or anything like that. I find some old things more beautiful than the new, and I continue to seek those places out, and I go back to them

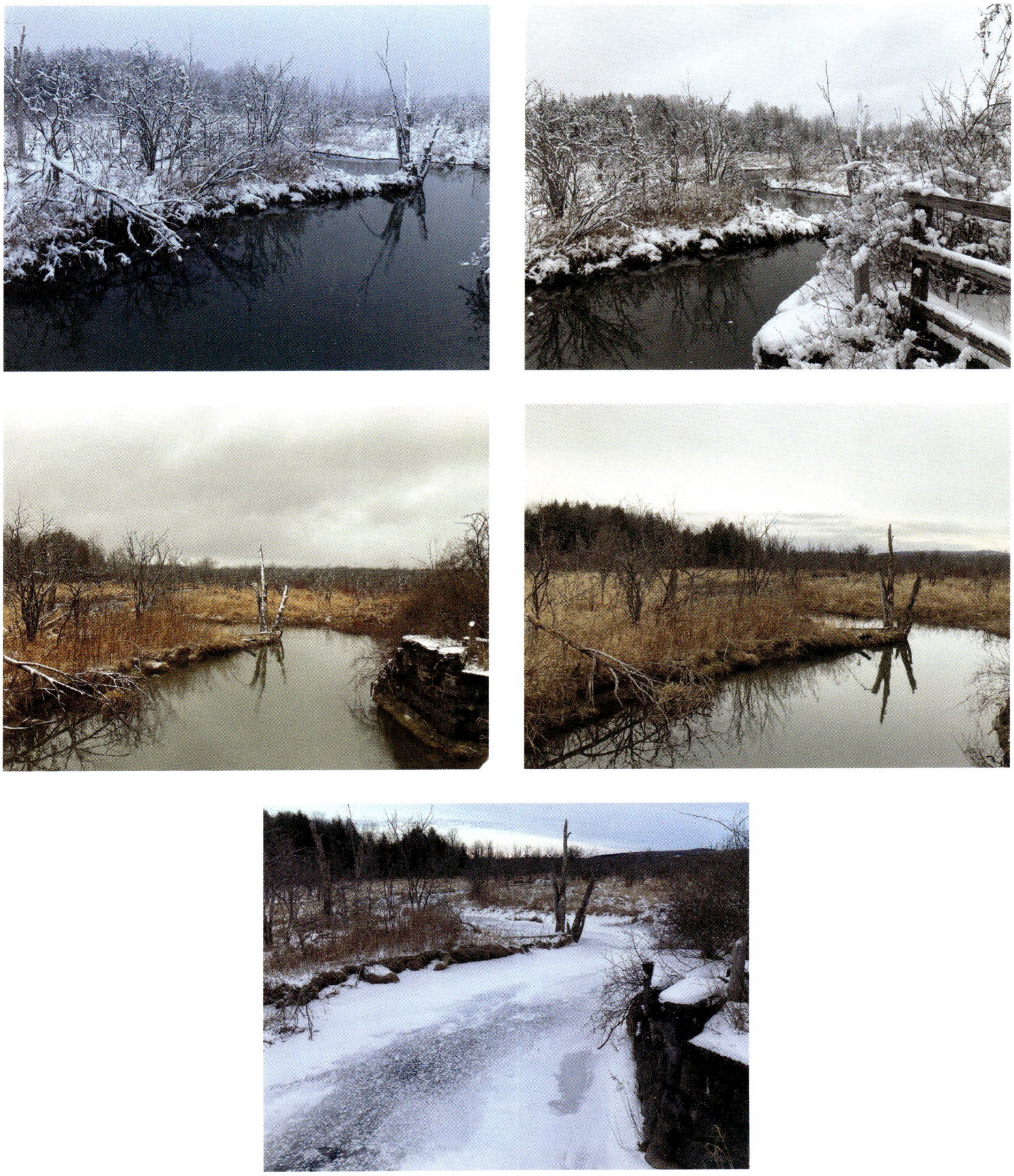

Figures 1–5
I have photographed this snag and bend in this stream on many winter walks. Each time I stop to photograph, I am interested in different phenomena having to do with the light and clouds, the color of the vegetation and the iced-over stream. My photographs are also influenced by the presence or absence of a friend and our dogs. The framings are different for each photo, yet they still show the snag and the bend in the stream, which I find to be memorable landscape features along this trail.

Images: Anne C Godfrey.

every year until sooner or later they are gone. They have blown away, burned, or fallen down, or just simply disappeared.[4]

Many of us maintain a similar relationship as the one Christenberry describes with a particular place or project (see Figures 1–5). These places may be designs we have made or, like for Christenberry, places that have great meaning to one's own personal history. Christenberry treated these photographs like touch points, traveling back and forth from his home in Washington, DC, to Hale County like one visits family. Photographs made in this way are interested in the dialogue between the photographer and the subject matter. Sometimes that dialogue is spoken but often not. Like any journey, there are many events and occurrences along the way to making the photographs that contextualize the experience of the photographer. Thus, mood, feeling and self-reflection may influence how a particular photograph is made on a particular day. This method allows for the imprint of the moment to come through. This is different than the more strict method discussed in Practice 4 "Re-photography," which maintains the same camera position as the first or original photograph.

CONCLUSION

This method acknowledges the interpersonal relationships that develop between designer and site, person and place. The concept of this method encourages the inevitable changes in this relationship (human and landscape) to manifest themselves by allowing the site to be photographed differently, yet still methodically, over time. The resulting photographs act as traces of this changing relationship, encouraging reflection and assessment when viewed together. In the end, this is another example of how the act of photographing, like the act of designing landscapes is much more than the sum of the resulting parts.

NOTES

1 Elizabeth Broun, William Christenberry, Howard Fox, Andy Grundberg, and Walter Hopps, *William Christenberry* (New York: Aperture, 2006).
2 William Ferris, "'Those Little Color Snapshots': William Christenberry," *Southern Cultures*, November 30, 2016, www.southerncultures.org/article/those-little-color-snapshots-william-christenberry/. This interview was recorded and filmed during the summer of 1983 at Christenberry's studio in Washington, DC.
3 For example, Christenberry created an ongoing body of work harshly criticizing the actions and cult of the Klu Klux Klan (KKK). His sculptural and painted work gets at the menacing qualities of the material culture the KKK used to reinforce its terrorist hold. This work of Christenberry's is difficult and often misunderstood without context. Christenberry states, "Some people have told me that this subject is not the proper concern of the artist or the art. On the contrary, I hold the position that there are times when an artist must examine and reveal such strange and secret brutality." *Memory Is a Strange Bell: The Art of William Christenberry* (New Orleans, LA: Ogden Museum of Southern Art, October 5, 2019–March 1, 2020), 5. Exhibition catalogue, accessed October 24, 2021, https://ogdenmuseum.org/exhibition/memory-is-a-strange-bell/.
See also Michael Adno, "Once It Comes Time," *Bitter Southerner*, accessed October 24, 2021, https://bittersoutherner.com/once-it-comes-time-william-christenberry-southern-photography.
4 Ferris, "'Those Little Color Snapshots'."

Practice 7 Joining: Combining

DISCUSSION

How, under the conditions of photography, can we start to represent the ephemeral qualities of landscape? As landscape architects and environmental designers, we know landscapes are dynamic and ever changing. How can we use photography to describe the passage of time and movement in landscape places?

This method introduces approaches for expanding the use of photography beyond the one-point perspective view of the single image. Here we will discuss practices of making multiple photographs of a landscape subject as a way to represent time, space, movement and the dynamic ever-changing characteristics of a place.[1] This approach to photographing also explores how cognitive experiences of the landscape can be portrayed.

Joining and combining multiple photographs becomes an accessible way to convey dynamism and change in the landscape. David Hockney's work serves as the primary influence here (to be discussed more extensively in the next section). Hockney developed his process for "joiners" in the 1980s as a way to confront the fact that the camera, with its single fixed lens, does not represent how we see.[2] The gaze of the lens is merely a simplified mechanical construction.[3]

Functionally, joining or combining represents a subject using more than one photograph. There are several options for the construction of these combinations, including displaying the multiple photographs in a grid[4] or into a nonlinear collage[5] or a photomontage.[6] These constructions allow for the representation of time passing and movement occurring. This is very different than how single photographic images freeze time and remove the moment from the continuum of constant change. Combined images disrupt our expectations for photography. Primarily because processes of the photographer become evident: the same hand appears in two or three different positions; multiple angles of an object appear together; the same animal appears in multiple areas of the scene; the light in the sky changes. These visual characteristics reveal the movement of the photographer and the passage of time necessary to make multiple photographs for these constructions. The interests of the photographer become evident. The photographer might dwell on one part of the larger subject more than another and thus makes more photographs of that particular part of the scene (see Figure 1).

Combining the photographs, either digitally or by hand, creates another psychological and temporal space to process both what is depicted in the photographs and to revisit

DOI: 10.4324/9780429284649-23

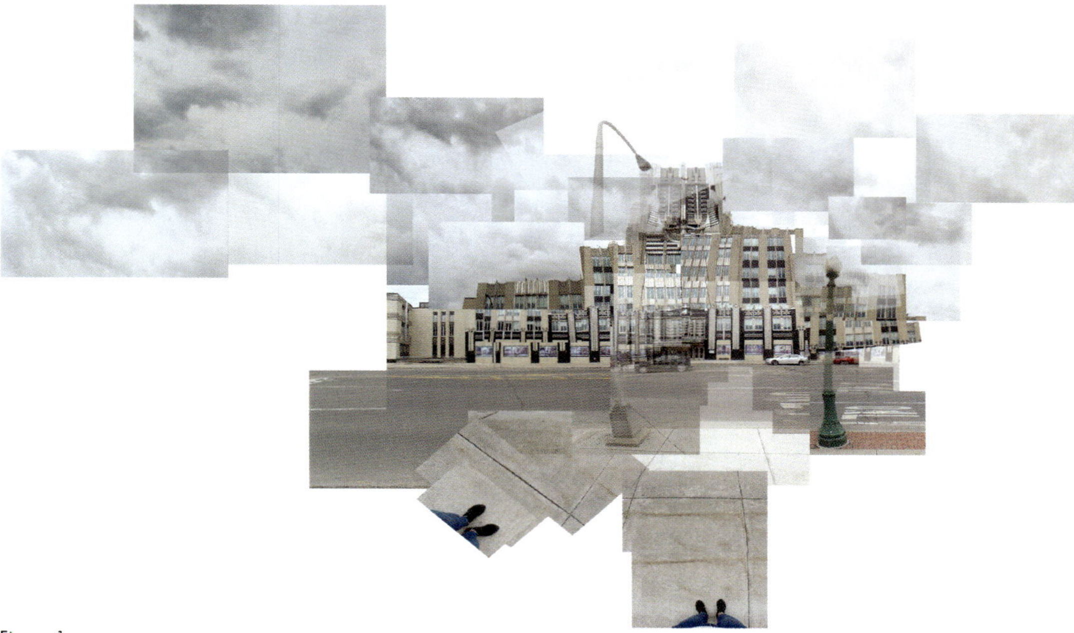

Figure 1
Niagara Mohawk Building, Syracuse, New York.
Image: Anna McGovern, March 2018.

the experience of photographing. Through reconstruction, the photographer/designer can remember the events and phenomena they experienced. All of those details may not be in the photographs themselves yet remain in the memory of the designer/photographer (and eventually play an influential role in design decision making). This process of constructing demands presence, attention and active noticing, which is different than, for instance, quickly flicking through a series of single images on a screen.

The construction of these collages or joiners is not meant to create seamless images; instead, the joints, spaces and near-matched connections are part of the image itself. Therefore, assembling these images is different than putting a puzzle together. These unmatched edges activate the eyes of the viewer. The obviously constructed collage qualities are curious.

For instance, the combination of edges – the edges of frames coming together – persuades the viewer to look differently. The eyes catch on these edges, stimulating more active eye movement around the image. A single subject matter may be present, but its articulation through multiple combined frames prevents the eye from resting quietly on that single subject. The edges of the frames stimulate dynamic eye movement, signaling the concept of dynamism in the landscape. Similarly, the appearance of the same subject multiple times (such as an animal moving across the field or a hand gesturing) or an object seen from multiple angles simultaneously is not what one expects, thus a kind of figuring out starts to happen. Hockney states, "You can only convince people that there is something beyond the conventionally made photograph if you alter it by making it even more vivid, more vividly real."[7] Though this way of working with photographs is no longer new, seeing joined photographs still piques attention. It is still not expected, thus stimulating curiosity and signaling that something should be looked at more closely. Finally, this process and result of combing are haptic. The resulting images indicate physical making as well as the physicality of the subjects.

INFLUENCES

Visual artist Tacita Dean, in her forward to David Hockney's 2018 monograph for the Richard Gray Gallery, writes, "A photograph is made up of one time: from the top left-hand corner to the bottom right, it is all the same fraction of a second." She continues explaining how Hockney found that fraction of a second never enough:

> For other artists, that ennui with the technological limitations of a medium might be enough to settle them into a lifetime working where they were comfortable, but David is insatiably curious. He set about finding a way to bring time into photography and began taking multiple Polaroids, then photographs, of a given subject and placing them side by side across the picture plane. Each photo-graph *[her hyphen]*, made its own time, then contextualizes time in the company of the next photograph: progression of taking, adjusting, and taking: the closest photography could get to his painting.[8]

Hockney's decades-long questioning of the static nature of individual photographs inspires him to break the physical and conceptual bounds of the single frame. Since the 1980s, Hockney has explored methods of combing multiple photographs together to transcend the "technological limitations" of photography. As the title of this 2018 monograph indicates, *Time and More, Space and More*, Hockney wants more from photography – more time, more space and more movement to occur in photographic representations.[9]

Hockney first explored arranging Polaroid photographs in a grid – making smaller images that collectively create a larger scene – such as *Yellow Guitar Still Life, Los Angeles, 3rd April 1982* and *Rain on the Pool, Los Angeles, April 12, 1982*. These "joiners" allowed for multiple perspectives and multiple scales to be shown. They could also include a longer amount of legible time. The making and construction of the joiners were part of the conceptual understanding of the work.

After these first prolific experiments, Hockney moved to roll film and expanded his explorations of combining multiple scales and points of view. *Pearlblosom Highway, 11–18th April 1982 (Second Version)* achieved what Hockney was trying to do – transcend photography with photography.

> Multiple viewpoints create a far bigger space than can be achieved by one. . . . I moved about the landscape, slowly constructing it from different viewpoints. The stop sign was taken head-on, indeed from a ladder, the words "stop ahead" on the ground were seen from above (using a tall ladder), and everything was brought together by "drawing" to create a feeling of wideness and depth, but at the same time everything was also brought up into the picture plane.[10]

Physically engaging a place with the multiple movements of photographing achieves an image representing the dynamic experience of the space. This addresses one of

Hockney's driving motivations (across all of his works) – taking on the problem of depiction. "Given that depicting, the urge to depict, is part of human nature, how do we go about it? What methods do we use and what assumptions do we make about depicting in this or that way?"[11] Hockney challenges the assumptions about how photography works – that it can only present still single moments with a single lens point of view. By simply combining two or more photographs together to depict a place, person or situation, he breaks the single point perspective of the camera, making an image closer to what we physiologically experience – that our eyes constantly move while we look to see any place or subject. Hockney came to this realization while working out the process for making *Walking in the Zen Garden at the Ryoanji Temple, Kyoto, February 21st, 1983.*

> I began to realize that one of the areas I was really examining was perspective, that this was what you could alter in photography. . . . To do it in photography was, in a sense, quite an achievement because photography is a picture making process totally dominated by perspective. Most photographers think that the rules of perspective are built into the very nature of photography, that it is not possible to change it at all. For me, it was a long process realizing that this does not have to be the case.[12]

Through his experiments with photography, Hockney presents a different way to see, experience and *think* about the subject. This process of photographing and constructing the joiners, as with any art practice, changes one's perception through the process of making. This idea, that the process of making alters and enriches our understanding of a place, is one of the benefits of working with this method. Hockney states more broadly, "In art, new ways of seeing mean new ways of feeling; you can't divorce the two, as, we are now aware, you cannot have time without space and space without time."[13]

Photographer Dawoud Bey also engages in the practice of combining photographs.

Staring with a 35mm camera for his early street portraiture in Harlem (1970s) and then moving to a Polaroid Type 55 film in a 4x5 format for his *Street Portraits*, Bey eventually began to collaborate with the Polaroid Artist Support Program and engage in studio work with a 20 x 24 Polaroid View Camera.[14]

His first work with the View Camera explored making diptychs of the same subject with different expressions or addresses to the camera, such as *Alva, New York, NY 1992*. This led Bey to "consider the possibility of representing a single individual through multiple images to reveal a more complex sense of character while extending the sense of time. . . . The idea of building the piece up, photograph by photograph, began to take hold as I continued to move past the convention of photographic singularity."[15]

Studio work with the Polaroid 20 x 24 View Camera is a time-intensive process, requiring the photographer, the camera technician and the subject to work together over hours of time.[16] The relationship between subject and photographer changes as this time passes in the making of the photographs. Instead of a taken moment, the resulting photographic image is the result of a cultivated relationship.[17] Bey states, "My growing notion was that photographs could indeed comprise multiple moments, allowing for the expansion of the single moment in photography to something representing time in the extended continuum in which the portrait is actually made."[18]

The process of making images, such as *Hillary and Taro, Chicago IL, 1992* and *Toyia, Kevin, Erica, Chicago IL, 1996*, results in seams across the faces where each of the 20 x 24 Polaroid images join. Like Hockney's work, these joints are not neat matches; instead, they show multiple movements and changes of expression in each face. The viewer is invited to linger, figure out, and look closely to explore the unique qualities of the individuals.

Bey more deeply explores the idea of description as the large camera is placed closer and closer to the subject, and multiple images are made to create a whole.

> Ultimately, the work expanded from diptych and triptych to monumental multi image works, moving the camera closer to the subject to create an even greater level of optical and material description, which then required additional photographs to fully describe the subjects. This formal experimentation was undertaken with the understanding that the end result would be compelling representations of young subjects that also advanced my evolving ideas of photographic interpretations and the photographic object itself.[19]

Each of the multi-image works shows movement, the passage of time and different facial expressions of the single or multiple human subjects. The edges, like with Hockey's work, activate the eye to look, compare and see what is there instead of function in a half-aware viewing as the mind fills in the spaces. Bey states, "This extended time is alluded to in the cinematic flickering of the subject as one reads across the various frames of the surface of the work itself."[20]

Sohei Nishino's *Diorama Maps* expand the working practices demonstrated by Hockney and Bey into massive wandering maplike photographic representations of whole cities, such as Kyoto or Rio de Janeiro. Nishino explores the city with his camera for weeks at a time, gathering impressions and memories through wandering and photographing.

After exploring, Nishino hand prints each black-and-white photograph and then begins the time-intensive manual construction process. These images become large, requiring ladders and expanding supports to accommodate their growth. When finished, the images read as recognizable places with aerial oblique photolike qualities – yet quickly, the bits and pieces of the details start to emerge. Multiple scales appear side by side, for instance, joined photographs of a whole building site next to a specific paving detail. Similarly, Nishino presents multiple viewpoints next to one another – an overhead view is next to a human-standing-eye-height view of a streetscape.[21]

Nishino discusses how his memories and impressions merge to guide his choices for construction.[22] The goal is not to create a kind of replica or even a legible map for wayfinding but to construct the experience of the place. Viewers look, tripping their own memory into action, as they stumble across a detail here or a building façade there. Looking at *Diorama Maps* becomes a give and take between Nishino and the viewer, the swirling constructions providing music to dance steps through the city and one's own memories.

More recently, the waves of the ocean became Nishino's photographic pallet because he was confined to his home village Mishima on the Izu Peninsula in Japan during the early stage of the COVID-19 pandemic. Starting in February 2020, Nishino began composing two new kinds of works. The first, made during the very beginning of the pandemic, is a collection of chaotic crashing waves. The second work, also of waves,

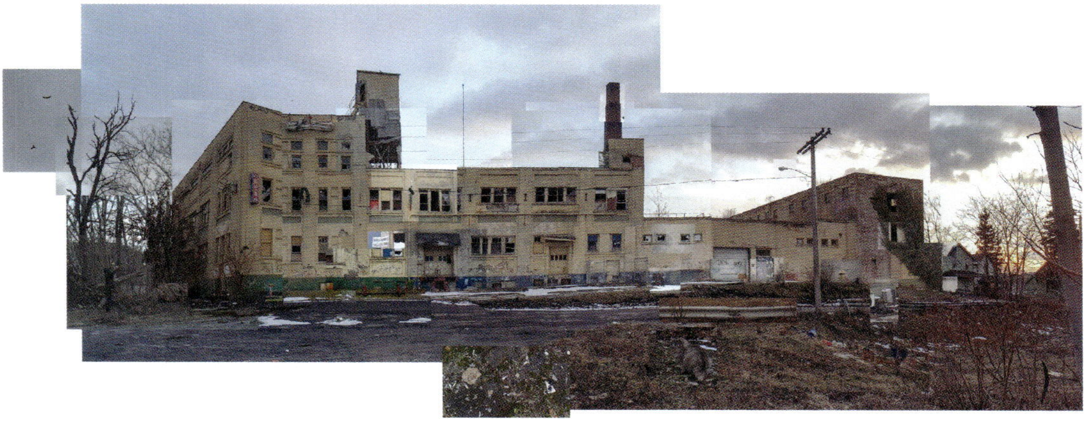

Figure 2
Danielle Ade combines different angles of view with enlarged details of interest. At first glance, this appears to be a cohesive image, but upon closer examination, the play of space and scale emerges.

Image: Danielle Ade, March 2018.

is calmer, reflecting time slowly passing as the pandemic continued on. These works are different than the *Diorama Maps* in that they have no underlayment of a mapped place to guide their construction. The waves, constantly moving, have no fixed points. Nishino had to interpret and construct from a different well of knowledge and experience. He says about this work, "It wasn't a reproduction or recording of my memory of a place. Instead, I'm creating this world."[23]

CONCLUSION

Combining and joining expands the processes of making and viewing into a more interactive relationship. Passage of time is required for both the photographing and the assembling (see Figure 2). The photographer must look deeply at the subject, sometimes calculating how the photographs will be made to later construct the combined image (such as Hockney does), while in other instances trusting a more open-ended process (like Nishino's work). (Re)assembling the photographs (be it physically or digitally) requires the photographer to look deeply at the subject matter again. During this reassembly process, memories are triggered. This could be anything from the sounds of traffic, wind or birds; sentences uttered to a colleague; notes made; soil samples collected; unexpected site features encountered; or unplanned interactions replayed.

NOTES

1 This discussion and underlying concepts were introduced in the first book of this series: *Active Landscape Photography: Theoretical Groundwork for Landscape Architecture* in Chapter 10. This process of combining or "joining" multiple photographs together is primarily inspired by the practices of David Hockney, who is discussed throughout this chapter.
2 See David Hockney, "Problems of Depiction: Moving Viewpoint," in *Hockney's Pictures* (New York: Thames & Hudson, 2016), 108–9.

3 When humans look and see, our two eyes (two lens sets) constantly move. The images projected onto the back of our eyes, our retinas, are conveyed to our brain and translated by a set of synaptic and neurological processes. This becomes a process of cognitive translation. We see with our brain. This is a much more complex and complicated process than what the camera does. The camera simply projects an image of a scene onto a light sensitive surface. The release of the shutter freezes a single moment onto the light-sensitive surface. This recording of this projection is controlled by (1) the type of lens (effected by such things as the shape, size and number or lenses housed within the lens case, optical coatings and use of filters) and (2) how, with digital photography, that image is processed by the on-board computer and a predetermined, preprogrammed algorithms that produce a set of aesthetics chosen by the programmers. These programmed presets influence color, contrast, dynamic range of highlights and dark tones and any number of other visual factors.
4 This is a direct application of David Hockney's first experiments with his "joiners" in which he arranged a set of Polaroid photographs in grid forms. For example, see David Hockney and Paul Joyce, *Hockney on Art* (London, England: Little, Brown, and Company, 1999).
5 After working with the Polaroids, Hockney moved to using a roll film camera. This lead to arranging the photographs in a more open-ended collage form that did not maintain rectilinear edges.
6 See, for example, Charles Waldheim and Andrea Hansen, *Composite Landscapes: Photomontage and Landscape Architecture* (Ostfildern, Germany: Hatje Cantz Verlag, 2014).
7 David Hockney and Nikos Stangos, *That's the Way I See It: David Hockney* (London: Thames and Hudson, 1993), 125.
8 Tacita Dean, "Forward," in *David Hockney, Time and More, Space and More . . .*, eds. Arthur Kolat, Tacita Dean, and David Hockney (Chicago, IL: Richard Gray Gallery, 2018), 7–8.
9 Arthur Kolat, Tacita Dean, and David Hockney, *David Hockney, Time and More, Space and More . . .* (Chicago, IL: Richard Gray Gallery, 2018).
10 Hockney, "Problems of Depiction," 108–9.
11 Ibid., 9.
12 Hockney and Stangos, *That's the Way I See It*, 100.
13 Hockney, "Problems of Depiction," 199.
14 Dawoud Bey, *Dawoud Bey: Seeing Deeply* (Austin, TX: University of Texas Press, 2018); Corey Keller and Elisabeth Sherman, *Dawoud Bey: Two American Projects* (New Haven, CT: Yale University Press, 2020).
15 Ibid., 136–37.
16 The camera is an example of a "mammoth" type camera with articulated bellows and the capacity to hold very large format film (in this case a Polaroid positive).
17 Also see Dawoud Bey and Brian Ulrich, *Dawoud Bey on Photographing People and Communities* (New York: Aperture, 2019).
18 Bey, *Dawud Bey: Seeing Deeply*, 137.
19 Ibid., 138.
20 Ibid.
21 *Sohei Nishino's maps trace more than the city* (September 3, 2018, San Francisco, CA: San Francisco Museum of Modern Art), film, 4:49, www.sfmoma.org/watch/sohei-nishinos-maps-trace-more-city/.
22 Sohei Nishino, interview by Alexander Strecker, "Diorama City Maps," *Lens Culture*, 2019, www.lensculture.com/articles/sohei-nishino-diorama-city-maps.
23 Sohei Nishino and Jon Mooallen, "The Art of Remaking Waves," *New York Times*, October 6, 2021, www.nytimes.com/interactive/2021/10/06/magazine/sohei-nishino-waves-photographs.html.

Practice 8 Rules

Games, Drifts, Prompts

DISCUSSION

Our experience of a landscape place, and our photographing of it, is mediated by a set of motivations, objectives and biases. Specifically, these mediations revolve around making design choices. This idea is a primary thread running through several of the discussions in this book. This method presents concepts from a diverse set of game-making practices that assist in relinquishing the influence of these mediations.

A game created through a set of rules is simple, easily executable and repeatable in any situation. This is different than Practice 2 "Predetermined Path." For that method, the specific landscape still influences what is circumnavigated or how the predetermined path unfolds. Here, the rules are stripped to the barest of parts of navigation. Walking or traversing. Turning. Following. Counting. The following prompt types and their examples will start to explain this method:

Counting and Turning

Every 10 feet/meters make one photograph looking forward and one looking back.

Take a pair of six-sided dice and roll them. Walk that many steps. Photograph in the four cardinal directions. Repeat.

Traverse 30 feet/meters, make a photograph looking forward and one looking back. Flip a coin to turn left or right. Then traverse another 30 feet/meters. Repeat.

After walking 20 steps turn to the left, then walk another 20 steps and then turn to the right. Repeat. Make a photograph at every turn in the direction of walking and then in the direction of turning.

A Ubiquitous Occurrence in a Landscape Prompts Action

At every red stop light or stop sign turn to the right. Photograph all four directions of the intersection.

As you approach an intersection controlled by a traffic light, go in the direction the pedestrian signal allows. Photograph looking ahead, up and down once you cross or turn.

Upon encountering a change of material in the ground plane, stop and photograph below, above and to either side. Then roll a dice or pull a card and walk as

DOI: 10.4324/9780429284649-24

many steps as the number indicates. If you encounter a change in the ground plane material before you finish the steps, stop, photograph as indicated earlier and then repeat with a new roll of dice or drawn card.

Movements Dictated by Other Peoples'and Animals' Actions in the Landscape

Follow someone wearing blue (or whatever color you choose) until you see another person wearing blue, then follow them. Repeat. While following, make a photograph every ten steps. If you lose your person, find another and start again.

While traversing a trail/sidewalk/path, make a panorama photograph set every time you hear a new bird call.

As you can see, anything is possible here, as long as you keep it simple. Some readers may already have experience with game theory, random number generators or dice play, while others may not. All of these rules are based on randomness. This method does not depend on any kind of expertise, skill or strategy.[1] Rules and games that become too complicated lead back to mediation of site experience because biases sneak back into the rules. For some readers, even the use of traffic signals may be too specific. In the end, you decide how far you take it.

Like the predetermined path or circumnavigation, implementation of an abstract rule or prompt takes the game player into parts of a site one may not have chosen to navigate otherwise. Without abstract rules, biases may lead the designer/photographer to focus on specific places while ignoring others. A rule removes the unique biases that can influence navigation through a site.

Taking a different turn, let's consider how this method may be applied to getting unstuck or unblocked. Let's say you are visiting a site for the fourth time in an attempt to figure something out. You know something is missing; your design is stalled; the inspiration is running dry; you're at the bottom of your bag of tricks. You might even be bored. Relinquishing control, through a toss of dice or pulling a card from a deck, might help to see things differently. Randomness can help us see anew. As does allowing for some playfulness.

Finally, no matter what kind of game is created, the only way to let it "work" is to allow yourself to do it. Another way to put this is you can't cheat. You must play the game. You have to buy into the underlying premise of the method, which is relinquishing control, in order to understand a place in a new way. This mindset, stimulated by the use of a game, must be maintained throughout the process. This takes a good deal of willingness to get uncomfortable. You may feel self-conscious, even conspicuous. If you are in a leadership role, this kind of game puts you on equal footing with anyone else playing the game.

The arbitrary nature of the rules does not mean that the experience remains arbitrary. You'll do things you wouldn't ordinarily do, and you will make photographs you would not ordinarily make.

INFLUENCES

Getting Unmediated

The approach to this set of methods is inspired in part by practices associated with psychogeography. The concepts and activities of psychogeography grew from an intermingling of concepts from Surrealism and Dadaism. Guy Debord first utilized the term "psychogeography" while undertaking a project to understand the city. He developed an interest in acknowledging the interrelationship between human emotions, memory and physical place. The Situationist International (SI), an outgrowth of these explorations, was founded by Debord and his contemporaries in 1957. This group formulated a set of practices for exploring the city through walking – called the dérive (or the drift).[2]

Over time, various veins of psychogeography have emerged. For the purposes here, the aspects involving game play as a serendipitous means to explore landscape places are of most interest. The SI thought of play as subversive and a way to unlock the unconsciousness. Their dérive practices, or drifting (as will be commonly referred to here), were originally developed for walking the city. Briefly, this idea of walking the city is partially inspired by the Flaneur. This figure, constructed by Charles Baudelaire, is a man who spends the day strolling and exploring the city. Within the context of Paris in the 1860s, the Flaneur, as depicted in *The Painter of Modern Life* (1863), is considered a signal for or symbol of modernity.[3] Susan Sontag, in *On Photography*, discusses how handheld, easily accessible cameras assist in the 20th-century practice of drifting: "The photographer is an armed version of the solitary walker reconnoitering, stalking, cruising the urban inferno, the voyeuristic stroller who discovers the city as a landscape of voluptuous extremes."[4] Into the late 20th century and early 21st century, contemporary practitioners of psychogeography have augmented the drift to explore nonurban spaces, such as agricultural lands, forests and suburbs – a departure from Debord's primary interests in the city.

Walking, or perambulation or traversing[5] remains at the core of the practice. The use of cars or other motorized vehicles that move quickly is considered a barrier to experiencing landscape place. Walking is valued because it is relatively slow (as compared to a train, motorbike or car) and also flexible, because the walker can change direction easily without a major interruption (not quickly possible in vehicular traffic, or while on a train when backtracking requires getting off one train and waiting to get on a return route). Traversing expands the possibilities to include singular human-scaled ways of navigating the landscape, such as rolling, that move at the same pace as walking and have similar flexibility.

Drifting is instigated by the creation of a set of arbitrary rules for navigating space. Those rules are outside of both the interests and desires of the drifter and the specific characteristics that make a place unique. By following these arbitrary rules, the drifter is untied from her mediated motivations for traversing.

This is all done as a means to see and discover anew. The drift is another example of how to get oneself into places that one may not ordinarily choose to go. There is a release of control, allowing the rule to dictate how one navigates a place. Our contemporary selves and lives may need exactly this – a way to release ourselves from constant mediation of experience through social media, agendas and constant work.

Getting Unstuck

As mentioned earlier, this practice of game play and premade prompts can assist in another common situation – being stuck. Often stuck-ness and its companion, burnout, indicate the overuse of the same techniques. Burnout is connected to a lack of space for experimentation and innovation. A balance of skill (utilizing past knowledge and experience) and challenge (introducing new factors and problems) combined with encountering new situations is crucial to cultivating creativity and flow.[6]

One way to get out of a rut and disrupt defaulting to the same well-worn tasks and moves is to use a deck of cards to randomly introduce new challenges. A deck of playing cards can be used by predetermining actions – each suit indicates and designates specific activities for face cards and number cards. Or a unique card deck can be custom-made with prompts.

The use of cards to interrupt the regular routine is not unique to psychogeography. Card play and prompts are used in multiple settings to restart the creative process. One contemporary example is Brian Eno's and Peter Schmidt's *Oblique Strategies*.

To get out of ruts in the recording studio, Eno, a groundbreaking producer and musician, created a set of cards with prompts. Eno first created this set for his own personal use. Around the same time, Eno's collaborator, Schmidt, also developed a set of prompt cards he titled *Thoughts Behind the Thoughts*. In 1974, they brought the two sets together to create a 115-card deck entitled *Oblique Strategies: Over One Hundred Worthwhile Dilemmas*[7] (see Figure 1).

Prompts include:

"Always the first steps."
"What mistakes did you make last time?"

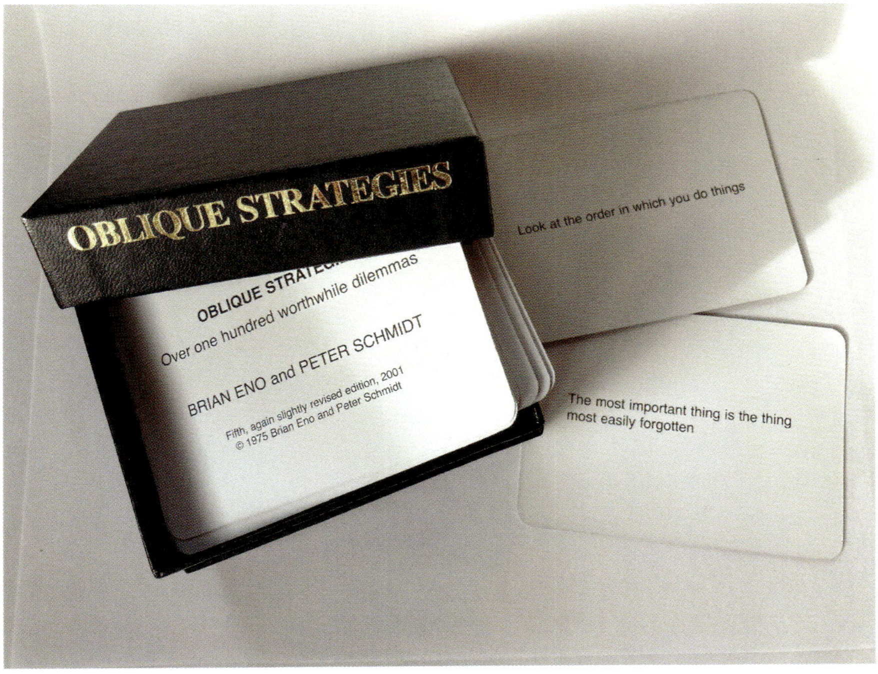

Figure 1
Some prompt cards from the *Oblique Strategies* set.

Image: Anne C Godfrey.

"The most important thing is the thing most easily forgotten."
"Think – inside the work – outside the work."
"Not building a wall, but making a brick."[8]

When working with musicians in the recording studio, Eno utilized the cards when the music-making process went stale. Some cards are more abstract and conceptual, asking a person to reassess their mindset, while others are very specific. Famously, one of the cards asks all of the musicians to switch instruments for part of the session – "Change instrument roles" – as a means to really shake things up by putting themselves in another person's fretboard or drumsticks. This was often frustrating, or even infuriating to some of the artists, but it led to breakthroughs. David Bowie's "Heroes" was produced using *Oblique Strategies*.[9]

Though born out of music making, the deck can be applied to any kind of creative endeavor. Thus, *Oblique Strategies* could be utilized as prompts while investigating and photographing a site.

CONCLUSION

These methods present intentional disruptions to the everyday practices we come to rely upon. They ask the photographer/designer to disengage from the expected and do the unexpected – in this case, relinquishing control over the process of navigating and photographing a site. This disruption opens up new avenues for discovery. It could assist in seeing what is missing, or perhaps see what is right in front of us, when otherwise predetermined expectations, agendas and biases veil our ability to notice them.

In another vein, the original spirit of playfulness, this open-ended exploration may very well lead to discoveries and experiences that remain unknown until they are happened upon and, in a sense, activated. This takes a good deal of willingness by the drifter to allow for exploring without a scripted agenda. It takes practice to cultivate this mindset – unbinding one's self from one's self. Photography, making new photographs while engaging in these playful prompts, becomes the process through which we translate our experiences. Photographing, along with the drifting, is the creative activity we use to re-see and re-understand a place.

NOTES

1 Which is quite unlike poker, where the player's success, to some degree, is assisted by skill and knowledge. For more on this see, Maria Konnikova, *The Biggest Bluff: How I Learned to Pay Attention, Master Myself, and Win* (New York: Penguin Press, 2020).
2 For example, see Francesco Careri, *Walkscapes: Walking as an Aesthetic Practice* (Ames, IA: Culicidae Architectural Press, 2017); Ken Knabb, *Situationist International Anthology* (Berkeley, CA: Bureau of Public Secrets, 2006).
3 Charles-Pierre Baudelaire, *The Painter of Modern Life* (London, UK: Penguin Books Ltd., 2010).
4 Susan Sontag, *On Photography* (New York: Farrar, Straus and Giroux, 2001), 55.
5 Traversing is used to here to acknowledge other human scaled singular ways of navigating the landscape besides walking, such as rolling. Contemporary discussion points out that the primacy of walking causally disregards those who are differently abled and may use some kind of motorization, wheels or other technology for navigating the landscape.

6 Mihaly Csikszentmihalyi, *Flow: The Psychology of Optimal Experience* (New York: Harper Row, 2009).

7 Maria Popova, "Oblique Strategies: Brian Eno's Prompts for Overcoming Creative Block, Inspired by John Cage." *Brain Pickings*, September 18, 2015, www.brainpickings.org/2014/01/22/brian-eno-visual-music-oblique-strategies/.

8 Brian Eno and Peter Schmidt, *Oblique Strategies: Over one Hundred Worthwhile Dilemmas*, V.5, Card Deck, 2001.

9 Tim Harford, *Messy: How to Be Creative and Resilient in a Tidy-Minded World* (London: Little, Brown Book Group, 2016), 7–32.

Practice 9 Many Others: Many Interpretations

DISCUSSION

This method focuses on collecting multiple interpretations of a place through looking at and talking about a selected set of photographs. Similar to Method 3 "Asking Questions," this method acknowledges individuals will share unique and diverse interpretations of the same photographs. This process creates a space for learning from the unique experiences of various stakeholders. To seek a broad set of interpretations, questions remain open:

> What do you see?
> What do you think about when you look at this photograph?
> How would you interpret this photograph?

The approach is interested in not overly influencing the responses by asking specific questions. For example, asking about a specific characteristic, *Could you tell me about this river, plaza, statue?*, directs the discussion into narrow territory. The respondent may look at a photograph that contains a statue, but their first inclination for interpretation of the photograph may or may not have anything to do with the statue. Thus, questions that are too specific may unintentionally negate the individual interests, memories or expertise of the viewer.

This method values that everyone thinks about photographs, or rather the content of the photographs, differently. Thus, how they choose to talk about a photograph will differ. One person might make a list of characteristics they see. Another person may relay a personal story from their childhood that references only one phenomenon in the photograph. Another person may share a piece of knowledge about a historical event. Another person may comment on specific aesthetic qualities of either the photograph or the subject matter (or some combination of both). Another person might talk about an ecological process. Another person may discuss the cost of something in the photograph. And most likely, each viewer will intertwine multiple types of information and experience in their interpretation.

The kind of information conveyed will vary. People's responses will be made through their own lenses or experience, expertise and value sets. Thus, a position of open curiosity defines the attitude of this method. Judgment is reserved. Sorting into "right and wrong" is put in check. This is a dialogue between designer and stakeholder that focuses on acknowledging the stakeholder's experience and thoughts, no matter how they respond or how "off topic" it might seem.

DOI: 10.4324/9780429284649-25

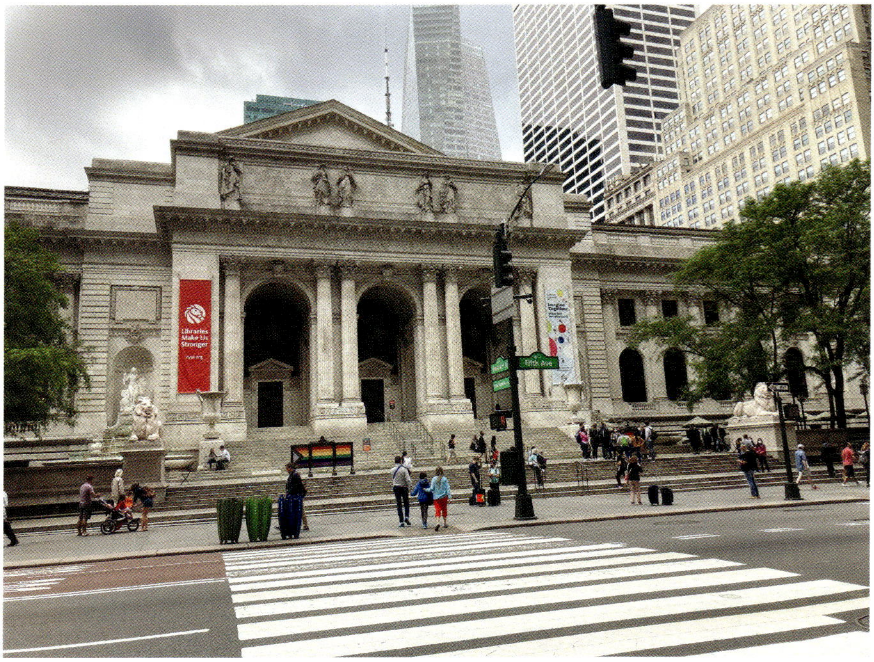

Figure 1
Keeping the questions open, such as, "Could you tell me what you think about when you look at this photo," allows for a diversity of responses. When looking at this picture of the New York Public Library, comments could vary. One person might convey a story about visiting the reading room as a child, another may discuss the history of the architecture, a third person recalls the man who would write short stories on a yellow typewriter while perched on the steps, another person would talk about the history of Bryant Park, which is behind the building.

Image: Anne C Godfrey, May 2021.

Two primary types of information arise in these discussions: trends and idiosyncratic interpretations. First, trends occur when a significant percentage of the viewers share certain types of knowledge consistently. This indicates a trend in understanding – be it cultural history, or present use or something that is concerning. Second, idiosyncratic interpretations often open up a door to knowledge that has been lost or erased. These outliers should not be dismissed but treated as indicators that something else is going on. What to do with this information may not be clear at first. Reserving judgment and letting that piece of knowledge continue to exist on its own merit may over time lead to understanding its significance.

By stepping out of our personal position, or that of our closest design team members' or even the inner circle of stakeholders', a larger and more diverse understanding of the site and its context can be cultivated through gathering. What might our design process look like if we engaged in gathering this broader set of interpretations?

This way of working is not novel. Community engagement is a core part of most public design projects. Design firms cultivate a specialty in engagement; municipalities require it; communities come to expect it. This chapter and method simply suggest a way to include a more expansive approach to the use of photographs – an approach that keeps the questions open enough to welcome a diversity of thoughts and experiences (see Figure 1).

INFLUENCES

The artist Sophie Calle's body of work *Take Care of Yourself*[1] serves as inspiration for this method. As with each of the practices in the second section of this book, this method extrapolates from a photographer's process, yet it is not meant to literally replicate it. The inspiration here lies in Calle's process of asking many different people from wide-ranging backgrounds and expertise to interpret something, in this case a piece of writing (instead of a photograph).

As part of the process, Calle makes a photograph of each woman who responds, adding it to their interpretation. Calle's many bodies of work combine text with photography (as well as sculpture and found objects sometimes). The text and photography enliven one another as they conceptually intertwine. In *Take Care of Yourself*, the text is a "breakup email" Calle received while traveling to China by train. The sign-off in the letter inspires the title of the work, "*take care of yourself.*"

In response, Calle asked a set of women, all experts in their various fields and occupations, to analyze and interpret the letter. Calle discusses this process of interpretation-by-others as a way to gain perspective on what was, at the time, a deeply personal and devastating event. "The idea came to me to develop an investigation through various women's professional vocabulary," Calle explains[2] (see Figure 2).

The experts are accomplished in a diverse set of professions – journalist, judge, physicist, anthropologist, lawyer, editor, intelligence officer, graphic designer. Some are artists – opera singer, comic, actress, performance artist, composer, rapper, screenwriter,

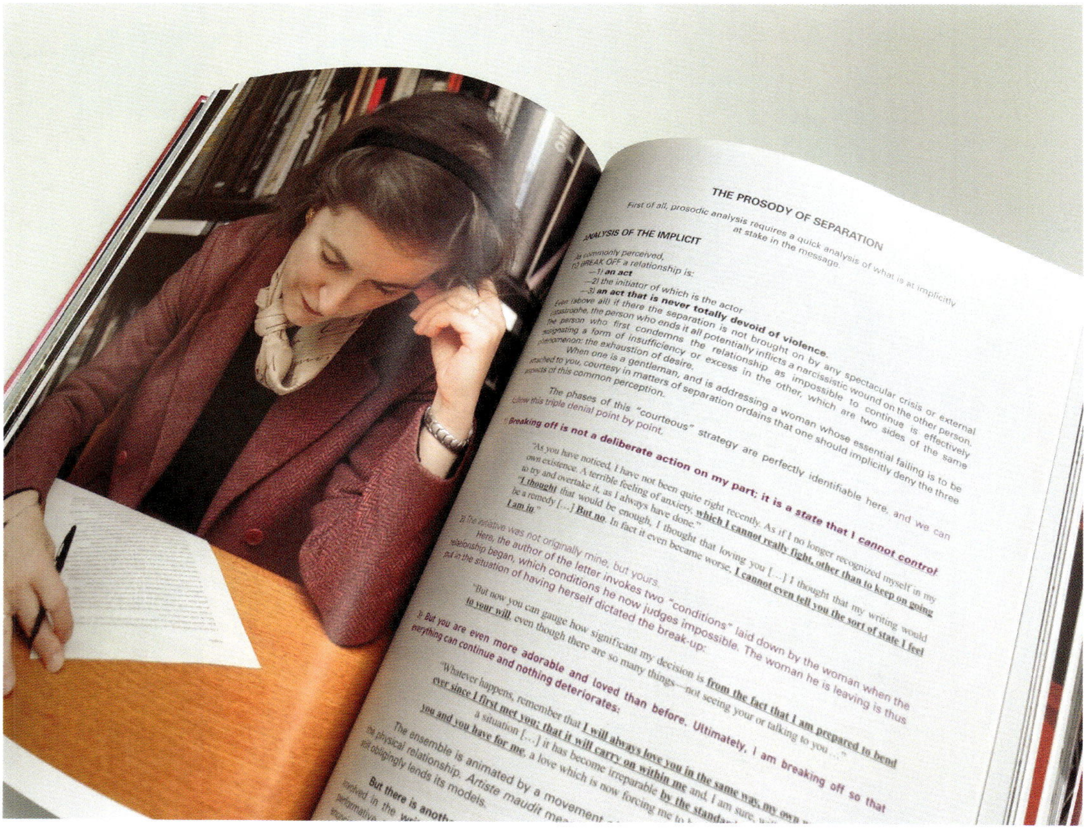

Figure 2
A stylistic analyst applies a prosodic approach to the reading of the letter. A photograph of *Take Care of Yourself*, Sophie Calle, 2007. *Image*: Anne C Godfrey.

clown. While others are experts in their different stages in life – grade schooler, teenager, elderly mother. Some of the women are friends or acquaintances, while others are people Calle was referred to by other interpreters.[3]

Calle articulates a requirement for the interpreters – that it must come from their position of expertise. It cannot be a sympathetic comfort for Calle's pain. Instead, she invites the interpretations to act as a filter or lens through which she can process and learn. She knows the event is too close to her. She wants to step away and see it from multiple viewpoints. Calle explains the process:

> The rules of the game are always very strict. In *Take Care of Yourself* I asked the participants to answer professionally, to analyze a breakup letter that I had received from a man. The parameters were fixed. For example, I wanted the grammarian to speak about grammar – I wanted to play with the dryness of professional vocabulary. I didn't want the women expressing sentiment for me. . . . I wanted to avoid any pathos or pathology.[4]

Though Calle's direction for interpretation, through professional expertise, is more specific than what is described in the opening to this method, her goal is similar. To collect a diversity of responses (see Figure 3).

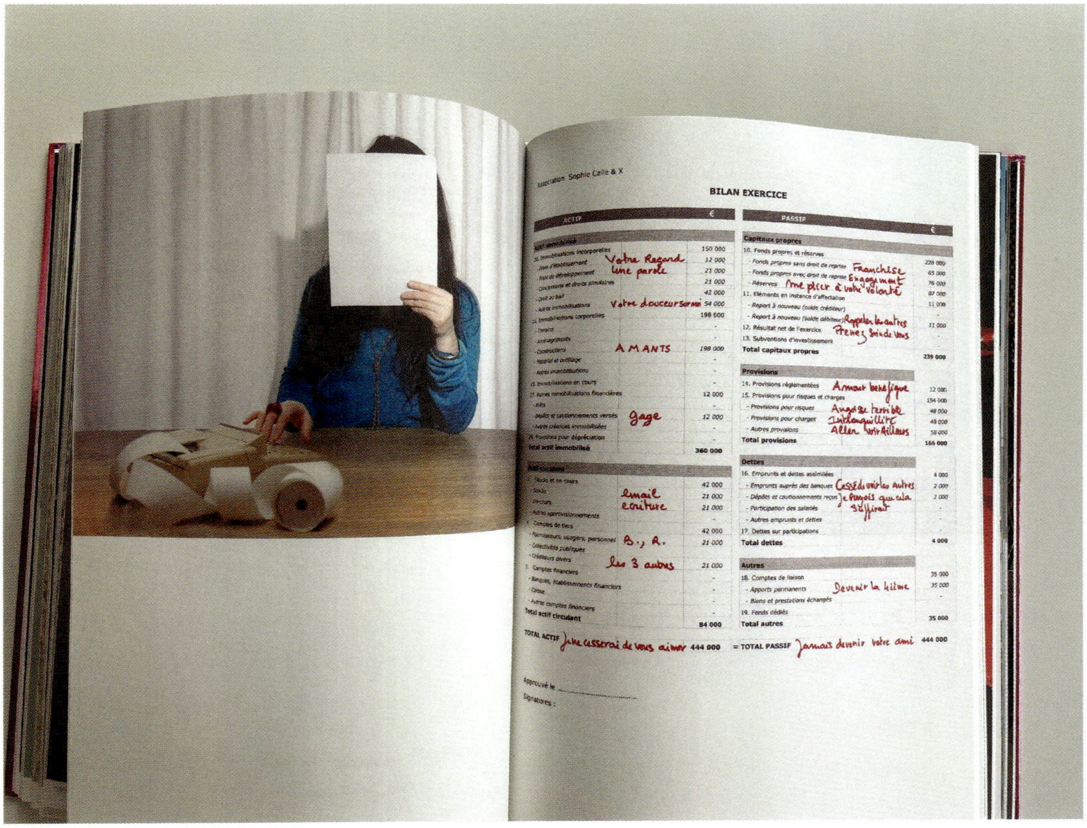

Figure 3
An accountant assesses the costs of the transaction. A photograph of *Take Care of Yourself*, Sophie Calle, 2007.

Image: Anne C Godfrey.

The text each expert produced is joined by either a photograph of her reading the letter or some kind of performance conveying her interpretation.[5] It is striking how many different interpretations occur when reading and viewing the body as a whole.[6] Many interpretations conflict with one another, while others collect into trends and shared sentiments – *the man is a narcissist; the man is compassionate; the man is trying to protect his own fragile sense of self; the man is callous; the man is in emotional turmoil.*

CONCLUSION

This method reminds us that individual people will have different understandings of a photograph (and text) based on their individual experiences and backgrounds. For our purposes, this understanding translates into acknowledging each person will have a different understanding of a landscape place. Though trends will emerge in the collection of these interpretations, the variety of unique knowledge and expertise, expressed through conversations or written words, can help environmental designers more deeply and holistically understand the unique characteristics of a place – be they physical, historical, cultural or systemic. A frame may be put on the process of interpretation, but as demonstrated with Calle's work, that frame need not constrain content, only the approach to interpretation. Could you please tell me what you think about when you look at this photograph? Could you tell me why?

Calle asks interpreters to stay within their discipline. This approach could be applied to our own work. Yet notice Calle asks the interpreters to respond to the letter *as a whole*. She does not say – what do you think of this sentence? What do you think of this word choice? So we may consider the same approach for our process – what do you think of this place?, rather than what do you think of this street, these trees, this lake?

A frame is created, but not too tightly. For example, one could ask a hydrologist to interpret a photograph based on their expertise, not on their personal history of that place. But to ask the hydrologist to talk about a specific feature presented in a larger scene becomes too directed for this method. One could ask a long-time resident to share their personal experience of this place, presented here in this photograph. But refrain from asking, What do you think of this playground, or this particular store front or this intersection?

Expertise can be articulated in many ways, from education, to cultural role, to codified job description. A community member could be a restaurant owner, a grade school educator, a computer programmer, an electrician, a freight company driver. A client could be an elected government official, a developer or a nonprofit environmental advocacy agency. This method helps in asking the unique individuals who are part of a project to share their specific ideas and expertise, pulling them out of a broader label of "neighbor," or "community member" or "client."

NOTES

1 *Take Care of Yourself* premiered in 2007 at the Venice Biennale in the French Pavilion and later toured across Europe and the United States. Artists are selected to display in their home country's Pavilion at the Biennale to acknowledge their influence and impact on art and culture in their country of origin.

2 Angelique Chrisafis, "He Loves Me Not," *The Guardian*, June 15, 2007, www.theguardian.com/world/2007/jun/16/artnews.art.
3 Sophie Calle, *Take Care of Yourself* (Arles: Actes Sud, 2007).
4 Sophie Calle, interview by Louise Neri, "Sophie Calle," Interview Magazine, March 7, 2009, www.interviewmagazine.com/art/sophie-calle.
5 Calle also filmed or recorded some of the women reading or interpreting in various ways, such as singing or dancing. Some of these recordings appeared at the premier of this work at the Venice Biennale in 2007, which I viewed. They are also collected in the book *Take Care of Yourself* on compact disks interspersed within the pages.
6 Either in its display, which I viewed at the Biennale in 2007, or in reading through the book.

Practice 10 Narrative

Writing What You Don't See

DISCUSSION AND INFLUENCES

Looking at photographs is quickly followed by constructing narratives about the subject matter. This method is interested in formally creating and combining written narratives with photography. These narratives are often complex, as they describe both visual characteristics in the photograph and other nonvisual characteristics impacting or informing the understanding of the subject matter. This method, for our uses in landscape architecture and environmental design, presents visual information within the photograph with non-visual information that influences the conditions of the subject matter.

Several methods in Part II: Practices acknowledge there are multiple points of view held about any landscape place. Thus, this method can stand alone or be combined with the previous methods. What makes this method different is that writing extensive and detailed narratives motivates the research and choices for photography, instead of the narratives being added at a later point.

There are many ways to present narratives with photographs. The following examples introduce several conventional and accessible approaches.

Long-Form Captions

Captions need not only name what is in the photograph. Captions can explain factors, histories and activities that are not seen but signaled by the subject matter of the photograph. The term "long form" refers to the actual length of these captions – a paragraph or more, instead of a sentence. This refers to long-form journalism, which, through its length, creates in-depth content addressing backstories, context and histories of the primary subject not possible in typical newspaper-story length journalism.

Richard Misrach, in several of his photographic bodies of work, pointedly utilizes long-form captions. These captions are not extra information; instead, they are an integral part of the photograph. Misrach is best known in the discipline of landscape architecture for his photographs in *Petrochemical America*, a collaboration with Aperture, Kate Orff and SCAPE.[1] In this collaboration, the environmental and cultural implications of oil extraction in the Gulf of Mexico are revealed through Misrach's photographs and SCAPE's complex analytical graphics. His photographs, the majority originally made in the 1990s, often include paragraph-long captions revealing histories not visually legible within the photograph itself. At times, this brings a particular point home by reuniting a past history with a present visual/physical situation. This unerases important events

DOI: 10.4324/9780429284649-26

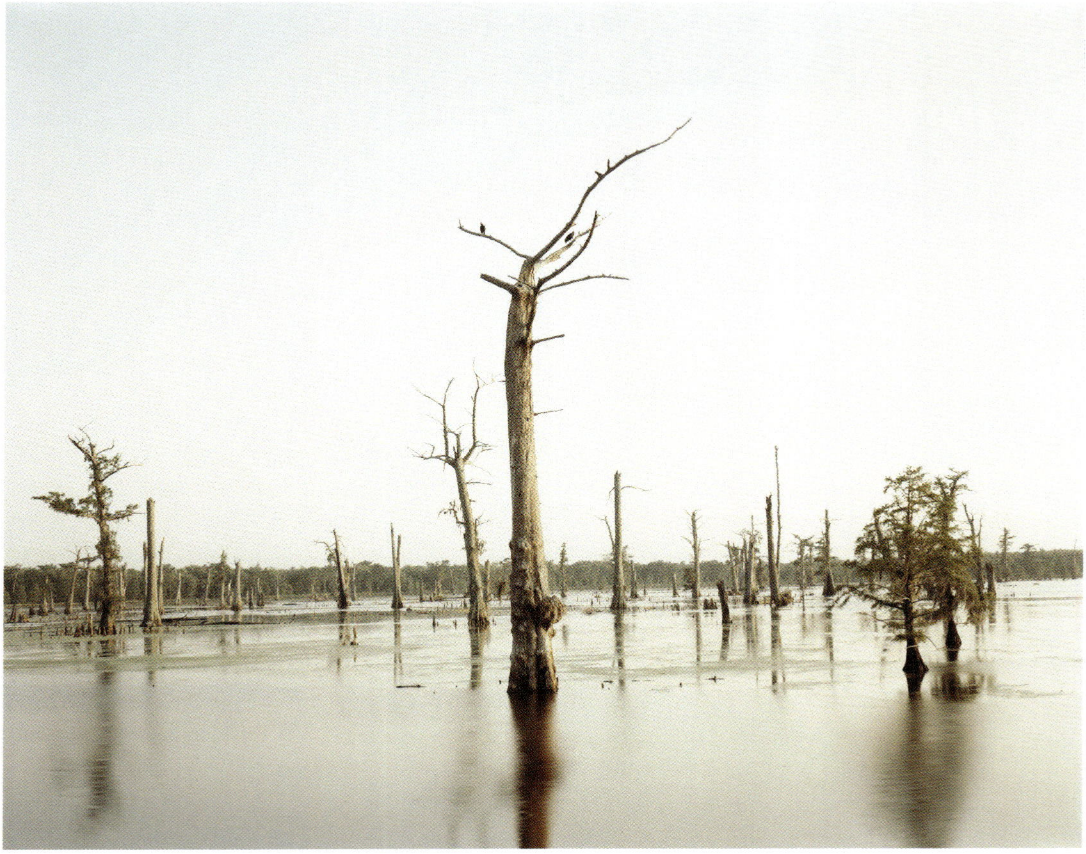

Figure 1
Cypress Swamp, Alligator Bayou, Prairieville, Louisiana 1998.

that have visually disappeared over time (see Figure 1). Misrach's photograph *Cypress Swamp, Alligator Bayou, Prairieville, Louisiana 1998* serves as one of several examples. The caption reads:

> Alligator Bayou is part of the Spanish Lake Basin, a fifteen-thousand-acre overflow swamp for a slow-moving body of water named Bayou Mancha. Several thousand years ago, Bayou Mancha was the main channel of the ever-changing Mississippi River. Native Americans thrived here thanks to its abundant natural resources. The twentieth century brought massive clear-cutting of cypress forests by lumber companies. In 1993, 1500 acres of lands and water rights were purchased to preserve the remaining wilderness and create an ecotourism business and educational center. Called Alligator Bayou, the area boasts 250 species of birds, free roaming alligators, and rare giant old growth Cypress trees. Subsequent sewage and human waste pollution from a nearby prison, followed by the timber industry's attempt to harvest the remaining Cypress trees, and most recently disputes with neighbors over water rights and water usage contributed to the closing of the wilderness area in 2010.[2]

Intertwining Narrative

Latoya Ruby Frazier's *The Notion of Family* combines the larger narrative of Black disenfranchisement in Braddock, Pennsylvania, with her family's personal narratives. Frazier presents photographs of her family next to photographs describing the environmental damages of steel and gas production in her hometown. Over time, the health of the Black and below poverty line inhabitants of the neighborhoods adjacent to the industrial sites has worsened. Yet the only accessible medical faculty in the area was shut down because it was deemed unprofitable.

> Seldom is there relief or aid for Black women living below the poverty line in our healthcare system. Tired of waiting in emergency rooms and being told by UPMC doctors that her migraines, chest pains, seizures, and shortness of breath are psychological, mom stays inside. She has become a prisoner in her home. Emotional fatigue has erected barriers around us.[3]

Intimate and unflinching, the photo narrative presents environmental and economic disenfranchisement in stark visual terms. These portrait and landscape photographs combine with Frazier's words, asking the viewer to truly see her mother, grandmother and family as real individual people with real agency, in a real place altered by calculated economic manipulation and disinvestment.[4]

Writing on the Photograph

Another approach is to write quite literally on the photograph itself. Done to disrupt the practice of floating the photographic rectangle in pristine white space, writing is applied to the photographic print itself or made part of the print in the dark room. At the time when all photographs were printed on light-sensitive, emulsion-coated paper, writing on the photographic print itself, beyond a signature, title or date was not considered normal practice, especially in art. Today, when photographs are presented in many different kinds of viewed spaces, from traditional prints, to digital outputs, to screens, this writing upon the photograph itself disrupts the notion of the photograph as a portal or window through which we look at an actual thing.

Duane Michals engages in this practice by applying his own handwriting to the white space around the rectangle of his photographic prints. The handwriting is slightly different and unique in each edition of these prints because Michals's hand writes each set. For his *Pittsburg* series, the photographs present a more recent version of Michals's home city, while his hand-written narratives reveal the past, telling of long forgotten events or things, like buildings, that are no longer in the photograph. Instead of acting as a window onto "what is there," Michals's work becomes a portal for multiple histories, with the photograph acting as the initiator of the narrative, not the result.[5]

In Michals's series *I Remember Pittsburg*, he writes his thoughts about the Alcoa Company and the Nixon Theater onto the white space of the photo paper (see Figure 2):

> The Nixon theater was a grand legitimate theater where I saw my first Broadway play. The balcony was so steep that one would look through one's knees to watch the production.

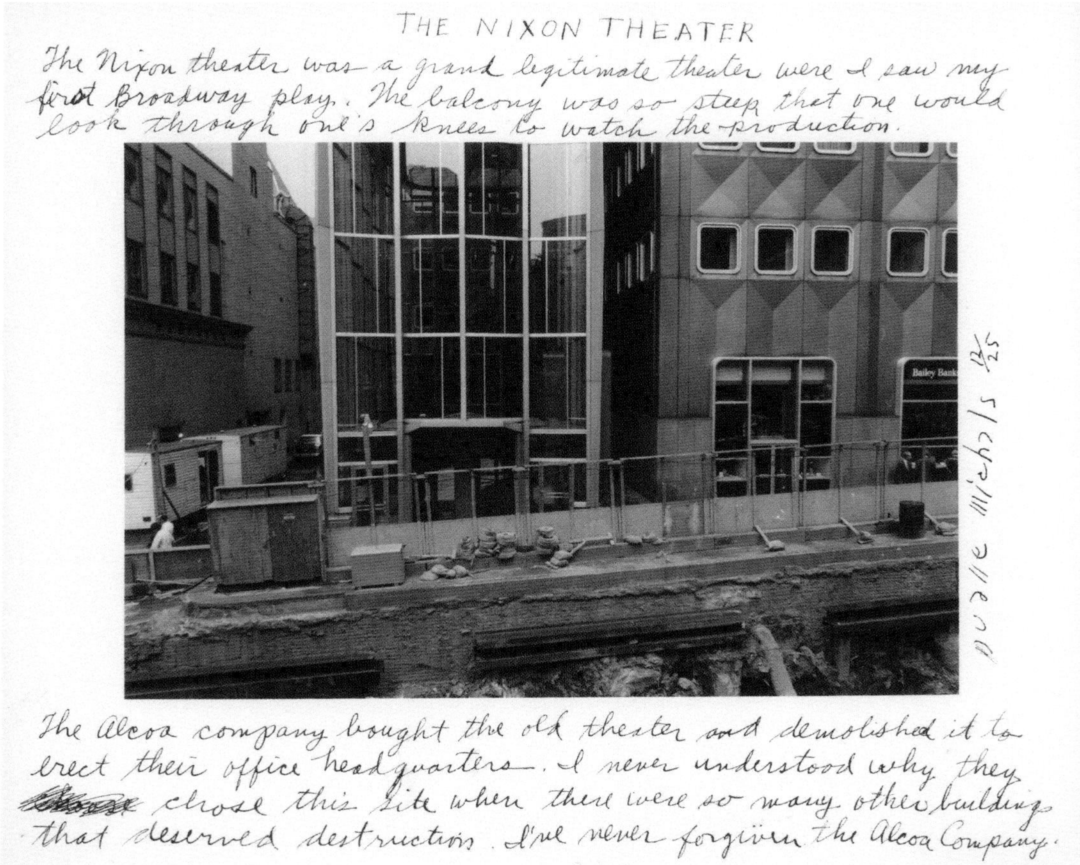

Figure 2
I Remember Pittsburgh, 1982.

> The Alcoa company bought the old theater and demolished it to erect their office headquarters. I never understood why they chose this site when there were so many other buildings that deserved destruction. I've never forgiven the Alcoa company.

Looking at the photograph, one sees the present condition. The narrative creates knowledge about a change in conditions. These conditions are physical – one building is there instead of another. These conditions are about use – a theater that served one purpose and now an office building serving another. And finally, the change in conditions is personal – Michals's own memories of attending the theater as a boy and his own thoughts, anger actually, at the destruction of the Nixon Theater by the Alcoa Company. There are no tidy divisions between these conditions. Michaels demonstrates the dynamic nature of looking at a photograph.

Collecting, Researching, Narrating

Kenneth Helphand, professor and author, has created several book projects that reveal complex cultural landscape histories by pairing photographs and researched narratives. One early work, *Colorado*, combines photographs from various sources, such as historic archives, photographers such as Robert Adams and Helphand's own collection, with an ongoing discussion about characteristics of the Colorado landscape.[6] Nearly every two-page spread of the book features a photograph next to Helphand's detailed narrative. The photographs do not serve as illustrations to the text, but rather stand equal to the text. Connective threads, such as the relationship between the cultural and environmental history of Colorado's arid landscape, clearly run through the work. As Helphand states in the introduction, "The interaction of people and place is embodied in the creation of the landscape."[7]

More recently, Helphand focused his attention on one unique cultural landscape phenomenon: the growing and harvest of hops in the Willamette Valley. The resulting book project *Hops: Historic Photographs of the Oregon Hopscape* presents photographs and narratives detailing the cultural life around the cultivation and picking of hops. Working closely with the Oregon Hops and Brewing Archive, Helphand collected photographs from holdings across Oregon, including the Independence Heritage Museum and the Oregon Historical Society, as well as images from private collections that had never been shared before.[8] As with *Colorado*, two-page spreads throughout the book present a single photograph joined by text. For *Hops*, the text combines Helphand's research and interviews with quoted descriptions from collected letters, newspaper articles and other historic documents. These descriptions, like Misrach's work, present discussion not only about what is presented in the photographs but also larger cultural histories about the long-standing traditions associated with the hop harvest. The viewer walks away with both a culturally nuanced and more technically aware understanding of hops cultivation.[9]

Annotations With Captions

Often used as a way to label items in a photograph, annotations (with leader lines, or numbers or some other kind of appended symbol) can become much more complex conveyors of information when concepts from long-form captions are applied to the annotation of photographs. For example, instead of simply pointing to and listing various species names, the annotations can give more complex information about, for example, what that species eats, specific behaviors or if it is common or uncommon to find the species in that place (see Figure 3).

In this case, we are also looking at annotations that are joined by a longer caption. The caption reinforces and contextualizes information given in the annotations. This creates different layers of entry into the photograph. Some viewers may only read the annotations, while others may also read the captions. A more active process of looking may occur, reading back and forth between the annotations and the caption, while also looking again and more closely at the photograph itself as the viewer takes in the information presented. This process instigates immediate learning – for instance, as I look at the visual characteristics presented here my understanding changes as I read the annotations

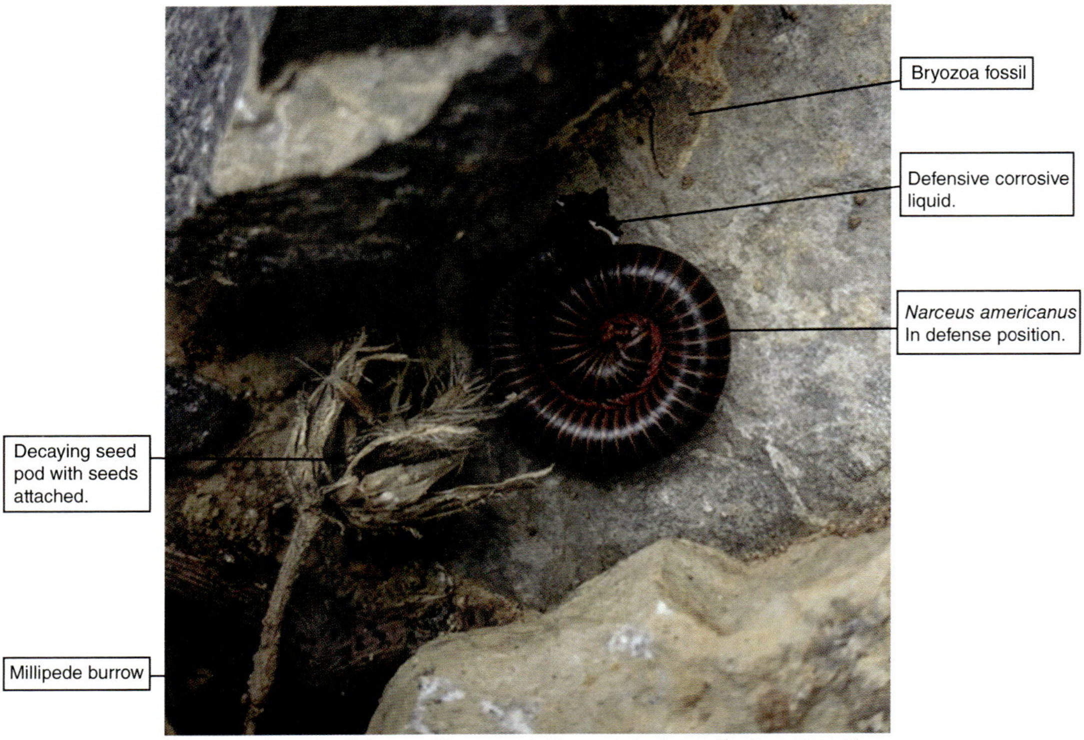

Narceus americanus in a defense position surrounded by a rocky landscape. The ***N. americanus*** is rolled in a ball as a defense mechanism to protect themselves from the photographer. The liquid located near the posterior end of the millipede is a corrosive liquid used to deter predators from digesting it. Next to the millipede is a decaying seed pod, presumably, the millipede was eating this seedpod prior to being disturbed by the photographer. They are a nocturnal species that spend most of their time in soil or rocky tunnels looking for and digesting organic matter. The tunnels are either made naturally or by a variety of organisms including the millipedes. They can house many different species from different species of millipedes to small vertebrates. This millipede was located on a five-foot shale hill where it was extremely hard to access. The location that the millipede was found in is not the right type of environment due to the lack of organic matter and suitable burrowing substrate.

Figure 3
Photograph labelled with annotations and leader lines.

Image: Brendan Cramphorn, October 2021.

and the caption. I don't just walk away with a species name and a visual impression, I understand multiple processes occurring within the image.

Annotations as a visual tool for communication are familiar – pointing at and labeling – thus, they are approachable. In some settings, they may be more approachable than long-form captions. A viewer can pick and choose through the annotations presented. They can read them in their order of interest, and in some situations, read them at their own pace. Annotations activate more visual interaction with the photograph as the viewer returns to look at each called out characteristic. This process prompts the viewer to make associations (see Figure 4).

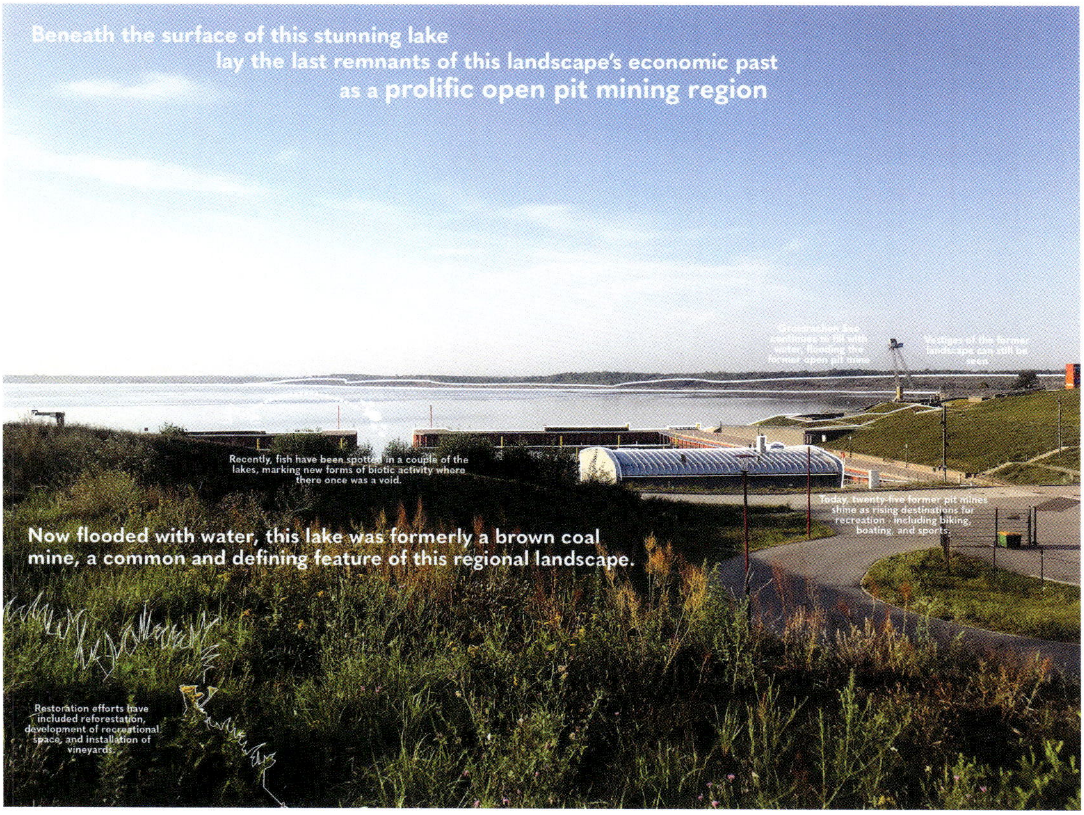

Figure 4
Photograph with annotations used as a visual tool.

Image: Anna Tiburzi, November 2021.

CONCLUSION

The practices in this final chapter return us to one of the main themes throughout this book: as environmental designers, we see and understand specific and multiple characteristics of a place in unique and specialized ways. Our training helps us explain these phenomena to others. Telling these complete stories by combining descriptive text (in this case) with photography, reveals complexity and presents new layers of understanding to our colleagues, peers, collaborators, stakeholders and community members.

For example, showing a picture of a lamprey in a presentation about Willamette Falls in Oregon may need no discussion if I am a member of the Confederated Tribes of Grand Ronde, or the cultural landscape historian for the project or someone who is familiar with Oregon's natural history.[10] But a picture of a lamprey alone, with a caption like "*Pacific lamprey*" or "Lamprey harvest, 2011" may be lost on a project engineer with an office in Boston, a developer from another state, or a concerned layperson who is not yet aware of this complex history. Instead, a caption could discuss the vulnerability of the lamprey to various impacts on the river system, from dams obstructing migration patterns, to pollutants negatively affecting the juvenal lamprey (called ammocoetes). The caption could also discuss that lamprey harvest is a significant cultural event for the Confederated Tribes of Grand Ronde and that lampreys are part of the foodways of the Tribes while serving medicinal and ceremonial purposes as well.

At face value, this method seems simple and is already practiced in many situations. Yet, telling these narratives influences the path of a project. Engaged parties can gain access to and share complex information with these methods. Landscape architects and environmental designers are investigators – we are also educators when we share information by putting into practice some of the examples presented in this chapter, and throughout this book.

Photography is a process of synthesis. It's a complex form of creation. And through active photography and the methods and practices in this book, we can add depth to our own understanding of any landscape.

NOTES

1 Richard Misrach and Kate Orff, *Petrochemical America* (New York: Aperture, 2014).
2 Ibid., 44–45.
3 LaToya Ruby Frazier et al., *LaToya Ruby Frazier: The Notion of Family* (New York: Aperture Foundation, 2016), 46.
4 Ibid., 149–53.
5 Duane Michals, *Storyteller: The Photographs of Duane Michals* (Munich: Prestel Verlag, 2014).
6 Kenneth Helphand and Ellen Manchester, *Colorado: Visions of an American Landscape* (Niwot, CO: Roberts Rinehart Publisher, 1991).
7 Ibid., xxi.
8 Kenneth Helphand, *Hops: Historic Photographs of the Oregon Hopscape* (Corvallis, OR: Oregon State University Press, 2020).
9 Ibid.
10 Office of Response and Restoration. National Oceanic and Atmospheric Administration, "With Restoration, Will Willamette River Lampreys Rebound for Northwest Tribes?" August 9, 2012, accessed September 3, 2018, https://response.restoration.noaa.gov/about/media/restoration-will-willamette-river-lampreys-rebound-northwest-tribes.html.

Acknowledgments

Preparing a book for academic publication involves far more than research and writing. I am thankful for three graduate assistants who have helped to bring this publication to life. First, Sara Constantineau worked with me this past year to organize images, manage spreadsheets, hunt down tricky citations, track permissions and provide a cheerful outlook when things got bogged down. She also contributed photography examples. Second, Anna Tiburzi patiently worked with me during the middle phase of this book in the thick of the COVID-19 pandemic. She managed many of the initial contacts for permissions, organized workflows and assisted initial case study research. Third, Kim Korioth was instrumental in managing the hand-in process for the first book in this series and worked with me during the initial phases of this book. The structures she created for image management, permissions tracking and citation guidelines have proven invaluable, and have been kindly passed down to Anna and then Sara.

I would not have been able to work with these excellent graduate assistants if it hadn't been for the generous support of the William Munsey Kennedy Jr. Fund provided to the Department of Landscape Architecture at SUNY College of Environmental Science and Forestry. Not only has the Kennedy Fund supported three graduate assistantships, but it has also provided funds for my own research. I am deeply grateful for the opportunities this funding has provided me over the last five years.

Thank you to Doug Johnston for valuing the meaning and impact of my teaching, research and book projects. Thank you to the Department of Landscape Architecture at SUNY College of Environmental Science and Forestry. My colleagues have given me the support, understanding and space that help an academic become successful in their research endeavors.

Thank you to Grace Harrison, my first editor at Routledge. She worked with me through the development and peer-review process for all three books in this series. Her thoughtful feedback combined with the recruitment of excellent peer reviewers has been instrumental to the development of this work. Thank you to the anonymous peer reviewers whose comments and challenges have assisted in refining major concepts across all of the books.

Thank you to Kate Schell and Megha Patel, my new stateside editorial team at Routledge for their support and patience with this work as the pandemic wore on. Thank you to Denise File at Apex CoVantange for managing the production of this book – it's far more complicated than it looks and I appreciate her behind the scenes management of all of the small yet important details.

Thank you as well to Routledge for supporting the growth of contemporary theory, diversity in academic voices and dissemination of social justice topics.

Many of the photo examples I am able to share throughout this book would not be possible without the enthusiastic students who have taken my photography classes at SUNY College of Environmental Science and Forestry over the last five years. Working with students in the expanded fields of environmental science and environmental studies has provided me with new insights for my ongoing research. It has been a pleasure to work with such a diverse set of students.

Colleagues and friends near and far continue to be supportive: Frankie Schweiger, Emma Froh, Wendy Andringa, Liz Deck, Arica Duhrkoop-Galas, Cindy Rice, Laurie Matthews and Romita Ray.

This book was written during the COVID-19 pandemic. I started the manuscript for this book on March 3, 2020, while on a retreat. The context of this time has a tremendous impact on every part of this book, from the photographers presented, to the concepts articulated, to the types of sources used.

What sustained me during this time were the countless lectures I attended, made possible by live streaming. I have been especially edified by lectures provided by the following institutions: Aperture, The Whitney, The International Center for Photography, The Eastman Museum, The Center for Creative Photography, Museum of Modern Art, San Francisco Museum of Modern Art, the New York City chapter of the American Society of Landscape Architects, the New York City chapter of the American Institute of Architects, the Design Trust for Public Space, Photoville, the Brooklyn Museum and David Zwirner Gallery. Also, the ongoing programming from Art 21 kept me inspired when everything felt dark.

I was also sustained by books. Piles of books arrived, informed, inspired and transported me during endless quiet days. Please buy books directly from the publisher, art institutions, your local bookstore and through Bookshop.org. This helps publishers and writers do the work they want to do.

This book is dedicated to Toni McCormick, who was a bright and colorful light in this rust belt city. Going to her office always felt like going home. She touched many lives in Central New York.

Finally, to my family Michael Godfrey, Rowan Godfrey and Jim Godfrey. To my partner in writing and life, Don Zolidis. To my co-pilot, Bardwell.

Bibliography

Adams, Robert. "Ecology." Art 21 video, 53:36 (minutes 27:00–38:00), November 11, 2007. https://art21.org/watch/art-in-the-twenty-first-century/s4/ecology/.

Adams, Robert. *Why People Photograph: Selected Essays and Reviews*. New York: Aperture, 2005.

Adno, Michael. "Once It Comes Time." *Bitter Southerner*. Accessed August 14, 2021. https://bittersoutherner.com/once-it-comes-time-william-christenberry-southern-photography.

Aikhsan, Wan Jerina Binti Wan. "High Line Competition." NYCSTUDIOARCH. Accessed July 17, 2020. https://cargocollective.com/Uofanycstudioarch/HIGH-LINE-COMPETITION.

Alderman, Naomi. *The Power*. New York: Back Bay Books Little, Brown and Company, 2017.

Als, Hilton. "An-My Lê." In *Aperture Conversations: 1985 to the Present*, edited by Melissa Harris and Michael Famighetti. New York: Aperture, 2018. Originally published in 2005.

American Museum of Natural History. *Worlds Beyond Earth*. 2020. New York. Film.

"Amy Sherald The Heart of the Matter." *Hauser and Wirth*. Accessed September 19, 2019. www.hauserwirth.com/hauser-wirth-exhibitions/24653-amy-sherald.

Aperture Foundation. "Aperture Conversations: Antwaun Sargent, Shikeith, and Naima Green on Black Queer Utopia." Youtube. Accessed February 19, 2021. www.youtube.com/watch?v=M8PiD2xRkuI.

Apple Inc. "Apple Introduces Dual Camera iPhone 11." Last modified September 10, 2019. www.apple.com/newsroom/2019/09/apple-introduces-dual-camera-iphone-11/.

Archigram. Accessed June 16, 2020. www.archigram.net/index.html.

Art in the Twenty First Century. Season 8, "Jeff Wall in 'Vancouver'." September 23, 2016, video, 13:28. https://art21.org/watch/art-in-the-twenty-first-century/s8/jeff-wall-in-vancouver-segment/.

"The August Sander Project." The Museum of Modern Art, November 17, 2016 – present.

"The August Sander Project." The Museum of Modern Art. Accessed March 16, 2021. www.moma.org/calendar/programs/113.

Baldwin, James. *If Beale Street Could Talk*. New York: Dial Press, 1974.

Balmori, Diana. *Drawing and Reinventing Landscape*. Chichester, West Sussex: Wiley, 2014.

Barasch, Alixandra, Kristin Diehl, Jackie Silverman, and Gal Zauberman. "Photographic Memory: The Effects of Volitional Photo-Taking on Memory for Visual and Auditory Aspects of an Experience." *Psychological Science* 28, no. 8 (June 2017): 1056–66.

Barcott, Bruce. "Making Airspace for Birds and Planes." *Audubon Magazine*, May–June 2009. Originally published in September-October 2008 as "Clearing the Air." www.audubon.org/magazine/may-june-2009/making-airspace-birds-and-planes.

Barrett, Terry. *Criticizing Photographs: An Introduction to Understanding Images*. Boston, MA: McGraw Hill Publishers, 2000.

Barthes, Roland. *Camera Lucida: Reflections on Photography*. New York: Hill and Wang, 2010. Originally published in 1981.

Barthes, Roland. *Elements of Semiology*. Translated by Annette Lavers and Colin Smith. London: Jonathan Cape, 1967. Originally published in 1964.

Bate, David. *Photography: The Key Concepts*. London: Bloomsbury Visual Arts, 2018.

Baudelaire, Charles-Pierre. *The Painter of Modern Life*. London, UK: Penguin Books Ltd., 2010. Originally published as *Le peintre de le vie Moderne* in 1863.

The BBC. "Biology KS3 & KS4: How Humans See." Accessed June 4, 2021. www.bbc.co.uk/teach/class-clips-video/biology-ks34-how-humans-see/z4wt382.

Benedict-Jones, Linda. "Duane Michals: Storyteller." Essay. In *Storyteller: The Photographs of Duane Michals*, 18–57. Munich: Prestel Verlag, 2014.

Benedict-Jones, Linda, Allen Ellenzweig, Aaron Schuman, Adam Ryan, and Marah Gubar. *Storyteller: The Photographs of Duane Michals*. Munich: Prestel Verlag, 2014.

Benjamin, Walter. *One-way Street and Other Writings*. London, England: Penguin, 2009.

Berger, Alan. *Reclaiming the American West*. New York: Princeton Architectural Press, 2002.

Bey, Dawoud. *Dawoud Bey: Seeing Deeply*. Austin, TX: University of Texas Press, 2018.

Bey, Dawoud. "Dawoud Bey on Visualizing History." SFMOMA video, 9:13, February 15 2020. www.sfmoma.org/exhibition/dawoud-bey-an-american-project/.

Bey, Dawoud. "Interview: Photographer Dawoud Bey on 'Class Pictures' and the Emory Project." By Rebecca Dimling Cochran. ArtsATL, July 20, 2010. www.artsatl.org/a-conversation-with-photographer-dawoud-bey-about-class-pictures-and-the-emory-project/.

Bey, Dawoud. "LaToya Ruby Frazier." In *Aperture Conversations: 1985 to the Present*, edited by Melissa Harris and Michael Famighetti. New York: Aperture, 2018. Originally published in 2014.

Bey, Dawoud, and Brian Ulrich. *Dawoud Bey on Photographing People and Communities*. New York: Aperture, 2019.

Bogost, Ian. "Your Phone Wasn't Built for the Apocalypse." *The Atlantic*, September 11, 2020. www.theatlantic.com/technology/archive/2020/09/camera-phone-wildfire-sky/616279/.

Broun, Elizabeth, William Christenberry, Howard Fox, Andy Grundberg, and Walter Hopps. *William Christenberry*. New York: Aperture, 2006.

Bulgakov, Mikhail. *The Master and Margarita*. London: Harvill Press, 1967.

Buse, Peter. *The Camera Does the Rest: How Polaroid Changed Photography*. Chicago, IL: University of Chicago Press, 2016.

Calle, Sophie. "Sophie Calle." By Louise Neri. *Interview Magazine*, March 7, 2009. www.interviewmagazine.com/art/sophie-calle.

Calle, Sophie. *Take Care of Yourself*. Arles: Actes Sud, 2007.

Campbell, Joseph. *The Hero with a Thousand Faces*. New York: Pantheon Books, 1949.

Canon USA. "Canon Cameras: From Beginner to Professional." Accessed August 24, 2021. https://shop.usa.canon.com/shop/en/catalog/cameras.

Careri, Francesco. *Walkscapes: Walking as an Aesthetic Practice*. Ames, IA: Culicidae Architectural Press, 2017.

Carson, Rachel. *Silent Spring*. Boston, MA: Houghton Mifflin Harcourt, 2002. Originally published in 1962.

Chrisafis, Angelique. "He Loves Me Not." *The Guardian*. Guardian News and Media, June 15, 2007. www.theguardian.com/world/2007/jun/16/artnews.art.

Christenberry, William. *House and Car, Near Akron, Alabama*. Twenty pigment prints. 1978–2005. Beverly Hills, CA: Marc Selwyn Fine Arts. www.marcselwynfineart.com/artists/william-christenberry/.

Christenberry, William. *Red Building in Forest, Hale County, Alabama*. Twenty pigment prints. 1974–2004. New York: Pace/MacGill Gallery. www.artsy.net/artwork/william-christenberry-red-building-in-forest-hale-county-alabama-2012.

Christenberry, William. *The Underground Club, Greensboro Alabama*. Twenty-one ektacolor prints. 1967–2005. New York: Pace/MacGill Gallery. www.phillips.com/detail/william-christenberry/NY040115/225.

Christenberry, William. *William Christenberry, Southern Photographs*. Millerton, NY: Aperture, 1983.

Christenberry, William, and Richard B. Woodward. *Kodachromes*. New York: Aperture, 2010.

Cicala, Roger. "The Camera Versus the Human Eye." PetaPixel, November 17, 2011. https://petapixel.com/2012/11/17/the-camera-versus-the-human-eye/.
Clarke, Graham. *The Photograph*. Oxford: Oxford University Press, 1997.
Coogler, Ryan dir. *Black Panther*. Burbank, CA: Marvel Studios, 2018. DVD.
Coppola, Francis Ford dir. *Apocalypse Now*. San Francisco, CA: American Zoetrope, 1979. DVD.
Cosgrove, Dennis. "Liminal Geometry and Elemental Landscape: Construction and Representation." In *Recovering Landscape*, edited by James Corner, 103–20. New York: Princeton Architectural Press, 1999.
Csikszentmihalyi, Mihaly. *Flow: The Psychology of Optimal Experience*. New York: Harper Row, 2009.
Cullen, Gordon. *Townscape*. London: The Architectural Press, 1968.
Curtis, James. *Mind's Eye, Mind's Truth: FSA Photography Reconsidered*. Philadelphia, PA: Temple University Press, 1989.
Daigle, Allain. "How the 50-mm Lens Became 'Normal'." *The Atlantic*, May 13, 2018. www.theatlantic.com/technology/archive/2018/05/how-the-50-mm-lens-became-normal/560276/.
Daston, Lorraine, and Peter Gailson. *Objectivity*. New York: Zone, 2010.
"David Hockney: Time and More, Space and More . . ." *Frieze*. Accessed June 28, 2019. https://frieze.com/media/david-hockney-time-and-more-space-and-more.
DC Moore Gallery. *Duane Michals: Mischievous Eye*. New York, November 14, 2019–January 4, 2020.
De Saussure, Ferdinand. *Course in General Linguistics*. Translated by Roy Harris. Chicago: Open Court, 1998. Originally published in 1916.
Deal, Joe. *West and West: Reimagining the Great Plains*. Chicago, IL: Center for American Places at Columbia College, 2009.
Deal, Joe, J. B. Jackson, Mark Johnstone, and Edward Leffingwell. *Joe Deal: Southern California Photographs, 1976–86*. Albuquerque, NM: University of New Mexico Press, 1992.
Deal, Joe, Mark Johnstone, Richard Meier, and Weston Naef. *Between Nature and Culture: Photographs of the Getty Center*. Los Angeles, CA: The J. Paul Getty Museum, 1999.
Dean, Tacita. "Forward." In *David Hockney, Time and More, Space and More . . .*, edited by Arthur Kolat, Tacita Dean, and David Hockney. Chicago, IL: Richard Gray Gallery, 2018.
"Design: Process and Tools." Guidance, Ministry of Housing, Communities and Local Government. Last Modified October 1, 2019. www.gov.uk/guidance/design.
Dingus, Rick. *The Photographic Artifacts of Timothy O'Sullivan*. Albuquerque, NM: University of New Mexico Press, 1982.
Doel, Ronald E., and Pamela M. Henson. "Reading Photographs: Photographs as Evidence in Writing the History of Modern Science." Essay. In *The Historiography of Contemporary Science, Technology, and Medicine*, 201–36. London, UK: Routledge, 2006.
Draplin, Aaron James. *Draplin Design Co.: Pretty Much Everything*. New York: Stewart, Tabori & Chang, 2016.
"Duane Michals." *DC Moore Gallery*. Accessed June 16, 2020. www.dcmooregallery.com/artists/duane-michals.
Dunklin, Thomas B. "Smith River Fish Count Instructional Video." *Smith River Alliance*, 13:28, August 6, 2009. Online video. https://vimeo.com/5972885.
Eastman Kodak Company. *Kodak Advertisement for the Brownie Camera*. 1900. Print, 4 ½ x 6 inches. Rochester, NY: Eastman Museum.
Eno, Brian, and Peter Schmidt. *Oblique Strategies: Over One Hundred Worthwhile Dilemmas*. V.5. Card Deck. 2001. Original edition published in 1975.
Eugenides, Jeffrey. *The Marriage Plot*. New York: Farrar Straus Giroux, 2011.
Eyre, Eric. *Death in Mud Lick: A Coal Country Fight Against the Drug Companies That Delivered the Opioid Epidemic*. New York: Scribner Book Company, 2020.
Ferris, William. "'Those Little Color Snapshots': William Christenberry." *Southern Cultures*, November 30, 2016. www.southerncultures.org/article/those-little-color-snapshots-william-christenberry/.

Fox, William L. *View Finder: Mark Klett, Photography, and the Reinvention of Landscape*. Albuquerque, NM: University of New Mexico Press, 2001.

Frazier, LaToya Ruby, Dawoud Bey, Dennis C. Dickerson, and Laura Wexler. *LaToya Ruby Frazier: The Notion of Family*. New York: Aperture Foundation, 2016.

Gaut, Berys, and Dominic McIver Lopes. *The Routledge Companion to Aesthetics*. London and New York: Routledge, 2001.

Gefter, Philip. "Quentin Bajac." In *Aperture Conversations: 1985 to the Present*, edited by Melissa Harris and Michael Famighetti. New York: Aperture, 2018. Originally published in 2013.

"Goal 1: Citizen Involvement." Oregon Planning Department of Land Conservation and Development. Accessed January 25, 2021. www.oregon.gov/lcd/OP/Pages/Goal-1.aspx.

Godfrey, Anne C. *Active Landscape Photography: Theoretical Groundwork for Landscape Architecture*. Abingdon, Oxon: Routledge, 2020.

Godfrey, Anne C. "Do We See What We See? More Than Icons for Landscape Photography." In *Active Landscape Photography: Theoretical Groundwork for Landscape Architecture*. London: Routledge, 2020.

Godfrey, Anne C. "Landscape Architecture and Photography: A Ubiquitous Relationship." In *Active Landscape Photography: Theoretical Groundwork for Landscape Architecture*. London: Routledge, 2020.

Godfrey, Anne C. "Making Photographs: Making Landscape." In *Active Landscape Photography: Theoretical Groundwork for Landscape Architecture*. London: Routledge, 2020.

Godfrey, Anne C. "Meaning Is a Relationship: Gathering and Photographs to Create Landscape Knowledge." In *Active Landscape Photography: Theoretical Groundwork for Landscape Architecture*. London: Routledge, 2020.

Godfrey, Anne C. "Photography for Environmental Design." Class lecture presented at SUNY College of Environmental Science and Forestry, Syracuse, New York, Spring Semester, 2018.

Godfrey, Anne C. "Photography, Environment and You." Class lecture presented at SUNY College of Environmental Science and Forestry, Syracuse, New York, Fall Semester, 2019 and 2020.

Godfrey, Anne C. "A Step Towards Multiplicity: Photography for Complexity." Conference Presentation at Council of Educators in Landscape Architecture (CELA) Annual Conference, Sacramento, CA, March 6–9, 2019.

Godfrey, Anne C. "Tanner Springs Moments." *Landscape Architecture Magazine*, February 2020.

Godfrey, Anne C., and John LePage. Interview with John LePage, Nelson Byrd Woltz. Personal, July 2018.

Guichard, Camille, dir. *Duane Michals: The Man Who Invented Himself*. New York: Alive Mind Cinema and Kino Lorber, 2015. DVD.

"Guidelines | HABS/HAER/HALS." National Park Service, April 30, 2016. www.nps.gov/hdp/standards/guidelines.htm.

Halprin, Lawrence. *The RSVP Cycles: Creative Processes in the Human Environment*. New York: Braziller, 1981.

Harford, Tim. *Messy: How to Be Creative and Resilient in a Tidy-Minded World*. London, UK: Little, Brown Book Group, 2016.

Harris, Melissa. "Sally Mann." In *Aperture Conversations: 1985 to the Present*, edited by Melissa Harris and Michael Famighetti. New York: Aperture, 2018. Originally published in 1995.

Harris, Melissa, and Michael Famighetti. "RoseLee Goldberg and Roxana Marcoci." In *Aperture Conversations: 1985 to the Present*, edited by Melissa Harris and Michael Famighetti. New York: Aperture, 2018. Originally published in 2015.

Hauser and Wirth Gallery. *Amy Sherald the Heart of the Matter*. New York, September 10, 2019–October 26, 2019.

Heiferman, M. *Photography Changes Everything*. New York: Aperture, 2012.

Heiferman, M., and S. Kelly. "The Look of Criminality." In *Seeing Science: How Photography Reveals the Universe*. New York: Aperture, 2019.

Helphand, Kenneth I. *Hops: Historic Photographs of the Oregon Hopscape*. Corvallis, OR: Oregon State University Press, 2020.

Helphand, Kenneth I., and Ellen Manchester. *Colorado: Visions of an American Landscape*. Niwot, CO: Roberts Rinehart Publishers, 1991.

Hess, Amanda. "Asian-American Actors Are Fighting for Visibility. They Will Not Be Ignored." *New York Times* (New York), May 25, 2016.

Hight, Eleanor M., and Gary D. Sampson. *Colonialist Photography: Imag(in)ing Race and Place*. London: Routledge, 2002.

"A History of Participatory Design." Make: Good, September 22, 2017. https://make-good.com/a-history-of-participatory-design/.

Hockney, David. *Hockney's Pictures*. New York: Thames & Hudson, 2016.

Hockney, David. "Problems of Depiction: Moving Viewpoint." In *Hockney's Pictures*, 108–9. New York: Thames & Hudson, 2016.

Hockney, David. "Time and More, Space and More . . . ." *Frieze Studios* film, 03:20.2018. https://frieze.com/media/david-hockney-time-and-more-space-and-more.

Hockney, David. "Way Out West: Space Exploration." In *A Bigger Message: Conservations with David Hockney*, edited by Martin Gayford, 143. London, England: Thames & Hudson, 2016. Originally published in 2009.

Hockney, David, and Paul Joyce. *Hockney on Art*. London: Little, Brown, and Company, 1999.

Hockney, David, and Nikos Stangos. *That's the Way I See It: David Hockney*. London: Thames and Hudson, 1993.

"ICONOCLAST Brief." Landscape Architecture Plus. Weitzman School of Design, University of Pennsylvania. Accessed June 16, 2020. https://laplusjournal.com/Brief.

Jackson, John Brinckerhoff, and Ervin H. Zube. *Landscapes: Selected Writings of J.B. Jackson*. Amherst, MA: University of Massachusetts Press, 1970.

Jazeel, Tariq. *Postcolonialism*. London: Routledge, 2019.

"Jeff Wall: Tattoos and Shadows." San Francisco Museum of Modern Art, March 2007, video, 2:21. www.sfmoma.org/watch/jeff-wall-i-begin-by-not-photographing/.

"Jeff Wall in 'Vancouver'." *Art21*. Accessed December 18, 2020. https://art21.org/watch/art-in-the-twenty-first-century/s8/jeff-wall-in-vancouver-segment/.

Jenkins, Barry, dir. *Moonlight*. New York: A24, Plan B Entertainment, and Pastel Productions, 2016. DVD.

Jenkins, William. *New Topographics: Photographs of a Man-Altered Landscape*. Rochester, NY: International Museum of Photography at the George Eastman House, 1975. Exhibition Catalogue.

"Joe Deal: West and West and Locating Landscape: Center for Creative Photography." University of Arizona Center for Creative Photography, April 13, 2010. https://ccp.arizona.edu/press/releases/joe-deal-west-and-west-and-locating-landscape.

Judd, D. *Donald Judd Spaces: Judd Foundation New York & Texas*. New York: Prestel, 2020.

Kahn, Eve M. "Yelling, Screaming, Almost Fisticuffs' . . . Over House Paint?" *New York Times*, March 7, 2019. www.nytimes.com/2019/03/07/style/sean-scully-gold-zinger.html.

Keller, Corey, and Elisabeth Sherman. *Dawoud Bey: Two American Projects*. New Haven, CT: Yale University Press, 2020.

Kelsey, Robin E. "Viewing the Archive: Timothy O'Sullivan's Photographs for the Wheeler Survey, 1871–74." *The Art Bulletin* 85, no. 4 (2003): 702–23.

Kimmelman, Michael. "A New $260 Million Park Floats on the Hudson. It's a Charmer." *New York Times*, May 20, 2021. www.nytimes.com/2021/05/20/arts/little-island-barry-diller.html.

Klett, Mark. Introduction to *Third Views, Second Sights: A Rephotographic Survey of the American West*, 3. By Mark Klett, Kyle Bajakian, William L. Fox, Michael Marshall, Toshi Ueshina, and Byron Wolfe. Santa Fe, CA: Museum of New Mexico Press, 2004.

Klett, Mark. "Rephotographing Nineteenth-Century Landscapes." Essay. In *Second View: The Rephotographic Survey Project*, 11–44. Albuquerque, NM: The University of New Mexico Press, 1984.

Klett, Mark. "Three Methods of Presenting Repeat Photographs." In *Repeat Photography: Methods and Applications in the Natural Sciences*, edited by Robert H. Webb, Diane E. Boyer, and Raymond M. Turner, 32–45. Washington, DC: Island Press, 2010.

Klett, Mark, Kyle Bajakian, William L. Fox, Michael Marshall, Toshi Ueshina, and Byron Wolfe. *Third Views, Second Sights: A Rephotographic Survey of the American West*. Santa Fe, CA: Museum of New Mexico Press in association with the Center for American Places, 2004.

Klett, Mark, Ellen Manchester, JoAnn Verburg, Gordon Bushaw, Rich Dingus, and Paul Berger. *Second View: The Rephotographic Survey Project*. Albuquerque, NM: The University of New Mexico Press, 1984.

Knabb, Ken. *Situationist International Anthology*. Berkeley, CA: Bureau of Public Secrets, 2006.

Kolat, Arthur, Tacita Dean, and David Hockney. *David Hockney, Time and More, Space and More . . .* Chicago, IL: Richard Gray Gallery, 2018.

Konnikova, Maria. *The Biggest Bluff: How I Learned to Pay Attention, Master Myself, and Win*. New York: Penguin Press, 2020.

Lakoff, George, and Mark Johnson. *Metaphors We Live By*. Chicago, IL: University of Chicago Press, 2011.

Lamb, Tom. "HALS Guidelines for Photography." *National Park Service*, July 2005. www.nps.gov/hdp/standards/HALS/HALSPhotographyGuidelines.pdf.

Lange, Susanne. *Bernd and Hilla Becher: Life and Work*. Cambridge, MA: MIT Press, 2006.

Latz, Amanda O. *Photovoice Research in Education and Beyond: A Practical Guide from Theory to Exhibition*. New York: Routledge, Taylor & Francis Group, 2017.

Link, Jeff. "Lethal Glass Landscapes." *Landscape Architecture Magazine*, February 2020. https://landscapearchitecturemagazine.org/2020/02/13/lethal-glass-landscapes/.

Low, Setha. *Spatializing Culture: The Ethnography of Space and Place*. London: Routledge, 2017.

Lynch, Kevin. *The Image of the City*. Cambridge, MA: MIT Press, 1979.

Marien, Mary Warner. *Photography and Its Critics: A Cultural History, 1839–1900*. Cambridge, England: Cambridge University Press, 1997.

Marien, Mary Warner. *Photography: A Cultural History*. New York: Harry N. Abrams, 2002.

Meinig, Donald William. *The Interpretation of Ordinary Landscapes: Geographical Essays*. New York: Oxford University Press, 1979.

Memory Is a Strange Bell: The Art of William Christenberry. New Orleans, LA: Ogden Museum of Southern Art, October 5, 2019–March 1, 2020. Exhibition Catalogue. Accessed August 14, 2021. https://ogdenmuseum.org/exhibition/memory-is-a-strange-bell/.

Michals, Duane. *ABCDuane: A Duane Michals Primer*. New York: Monacelli Press, 2014.

Michals, Duane. *Duane Michals: Portraits*. New York: Thames and Hudson, 2017.

Michals, Duane. "This Is a Real Dream." Interview by William Jenkins, in *Storyteller: The Photographs of Duane Michals*, 84–88. Munich: Prestel Verlag, 2014. Originally published in *The Photographer's Choice: A Book of Portfolios and Critical Opinions*, 35–44. Danbury, NH: Addison House, 1975.

Misrach, Richard. *Richard Misrach on Landscape and Meaning*. New York: Aperture, 2020.

Misrach, Richard, and Kate Orff. *Petrochemical America*. New York: Aperture, 2014.

Misrach, Richard, and Myriam Weisang Misrach. *Bravo 20: The Bombing of the American West*. Baltimore, MD: Johns Hopkins University Press, 1990.

Mook, Bobby. *FOMO – The Fear of Missing Out: Bobby Mook at TEDxUNC*. TEDx Talks, March 20, 2014, video, 8:44. www.youtube.com/watch?v=1mZAQC9djPE.

Moore, Kathryn. *Overlooking the Visual: Demystifying the Art of Design*. Abingdon, England: Routledge, 2010.

The Morgan Library & Museum. "Illusions of the Photographer: Duane Michals at the Morgan." The Morgan Library & Museum, 4:19, December 19, 2019. www.themorgan.org/exhibitions/duane-michals.

Moyers, Bill, dir. *Joseph Campbell and the Power of Myth*. Aired on June 21 1988 on PBS. www.amazon.com/gp/video/detail/amzn1.dv.gti.1cb1064f-c5ce-dfc7-2e1b-083b80168d1a?ref_=imdbref_tt_wbr_pvt_aiv&tag=imdbtag_tt_wbr_pvt_aiv-20.

Mullaney, Jennette. "Looking Back at 15 Years of the High Line." *The High Line*, June 9, 2014. www.thehighline.org/blog/2014/06/09/looking-back-at-15-years-of-the-high-line/.

Museum of Modern Art. Dorothea Lange. Words and Pictures. New York, February 9–September 19, 2020. Online. www.moma.org/calendar/exhibitions/5079.

Museum of Modern Art. *The Family of Man*. New York, January 24–May 8, 1955.

Museum of Modern Art. Stephen Shore. New York, November 19, 2017–May 28, 2018.

National Park Service. "Heritage Documentation Programs: HABS, HAER, HALS." National Park Service, October 21, 2021. www.nps.gov/hdp/.

Nelson, Maggie. "Catherine Opie." In *Aperture Conversations: 1985 to the Present*, edited by Melissa Harris and Michael Famighetti. New York: Aperture, 2018. Originally published in 2017.

Neri, Louise. "Sophie Calle." *Interview Magazine*, March 7, 2009. www.interviewmagazine.com/art/sophie-calle.

"New York State Citizen Participation Plan." Office of Policy and Strategic Plan, New York State Homes and Community Renewal, May 20, 2020. https://hcr.ny.gov ›citizen-participation-plan.

Newport, Cal. *Digital Minimalism: On Living Better with Less Technology*. London: Penguin Business, 2020.

Nishino, Sohei. "Diorama City Maps." By Alexander Strecker. Lens Culture, 2019. www.lensculture.com/articles/sohei-nishino-diorama-city-maps.

Nishino, Sohei, and Jon Mooallen. "The Art of Remaking Waves." *New York Times*, October 6, 2021. www.nytimes.com/interactive/2021/10/06/magazine/sohei-nishino-waves-photographs.html.

Nolan, Christopher, dir. *Memento*. 2000; Santa Monica, CA: Summit Entertainment, 2001. DVD.

Nordstrom, Alison. "After New: Thinking About New Topographics from 1975 to the Present." In *New Topographics*, edited by Britt Salvesen, 69–79. Göttingen: Steidl, 2010.

Office of Response and Restoration. "With Restoration, Will Willamette River Lampreys Rebound for Northwest Tribes?" National Oceanic and Atmospheric Administration, August 9, 2012. https://response.restoration.noaa.gov/about/media/restoration-will-willamette-river-lampreys-rebound-northwest-tribes.html.

Palmer, James F., and Robert Sullivan. "Visual Prominence as Perceived in Photographs and In-Situ." *Journal of Digital Landscape Architecture* 5, no. 1 (May 2020): 286–94.

Palmer, James F., Vincent Vanderheyden, Gisele Alves, and Georgia Sismani. "Best Focal Length to Represent a Landscape View Using a Single-Frame Photograph." *Journal of Digital Landscape Architecture* 2, no. 1 (2017): 236–43.

Pier55 Inc. "Little Island." Last modified May 31, 2021. https://littleisland.org/.

"Planning and Development." The New Mexico State Land Office. Accessed January 25, 2021. www.nmstatelands.org/divisions/commercial-resources/planning-and-development/.

Popova, Maria. "Oblique Strategies: Brian Eno's Prompts for Overcoming Creative Block, Inspired by John Cage." *Brain Pickings*, September 18, 2015. www.brainpickings.org/2014/01/22/brian-eno-visual-music-oblique-strategies/.

"Prepare for Your Visit." *Kentuck Knob*. Accessed June 4, 2021. https://kentuckknob.com/visit/prepare-for-your-visit/.

Rion, Gilles. "Guy Debord: The Naked City, 1957." Frac Centre-Val de Loire, July 4, 2016. www.frac-centre.fr/_en/art-and-architecture-collection/debord-guy/the-naked-city-317.html.

Risko, Evan F., and Sam J. Gilbert. "Cognitive Offloading." *Trends in Cognitive Sciences* 20, no. 9 (August 2016): 676–88.

Robertson, Rebecca. "Duane Michals: Fighting Against Photography." *ARTnews.com*, July 29, 2013. www.artnews.com/art-news/artists/duane-michals-fighting-against-photography-2274/.

Rosenblatt, Louise M. *Literature as Exploration*. New York: The Modern Language Association of America, 1995. Originally published in 1938.

Rouvalis, Cristina. "Duane Michals: Telling the Story of the Storyteller – Carnegie Museum of Art." *Storyboard*, October 28, 2014. https://storyboard.cmoa.org/2014/10/duane-michals-telling-the-story-of-the-storyteller/.

Ruscha, Edward. *Twentysix Gasoline Stations*. Alhambra, CA: Cunningham Press, 1969.

Salvesen, Britt. *New Topographics*. Göttingen: Steidl, 2010.

Salvesen, Britt. "New Topographics." In *New Topographics*, edited by Britt Salvesen, 11–67. Göttingen: Steidl, 2010.

Salvesen, Britt. "'Surrealistic and Disturbing': Timothy O'Sullivan as Seen by Ansel Adams in the 1930s." *Journal of Surrealism and the Americas* 2, no. 2 (2008): 162–79.

Samsung. "How to Use the Camera on a Galaxy Smartphone." Accessed August 27, 2021. www.samsung.com/uk/support/mobile-devices/how-to-use-the-camera-on-a-galaxy-smartphone/.

Sandweiss, Martha A. *Print the Legend: Photography and the American West*. New Haven, CT: Yale University Press, 2002.

Sargent, Antwaun. "The Future Will See You Now." *Aperture 241 Utopia*, December 8, 2020.

Schiff, Richard. "Add Red, or Not." In *Joan Mitchell*, edited by Sarah Roberts and Katy Siegel. New Haven: Yale University Press, 2021.

Scoates, Christopher, Brian Eno, Roy Ascott, Steve Dietz, Brian Dillon, and Will Wright. *Brian Eno: Visual Music*. San Francisco, CA: Chronicle Books, 2013.

"Scout." *Merriam Webster*. Accessed April 14, 2021. www.merriam-webster.com/dictionary/scout.

Sherald, Amy, and Kehinde Wiley. "Brooklyn Talks: Amy Sherald and Kehinde Wiley." By Doreen St. Felix. Brooklyn Museum, October 14, 2021. www.brooklynmuseum.org/calendar/event/brooklyn_talks_sherald_wiley_october_2021/.

Shikeith, and Naima Green. "Aperture Conversations: Antwaun Sargent, Shikeith, and Naima Green on Black Queer Utopia." By Antwaun Sargent. Aperture Foundation, February 19, 2021. www.youtube.com/watch?v=M8PiD2xRkul.

Shore, Robert. *Post-Photography: The Artist with a Camera*. London, England: Laurence King Publishing, 2014.

Shore, Stephen. *The Nature of Photographs*. Baltimore: The Johns Hopkins University Press, 1998.

Snyder, Blake. *Save the Cat! The Last Book On Screenwriting You'll Ever Need*. Studio City, CA: M. Wiese Productions, 2005.

Soares, Julia, and Benjamin C. Storm. "Forget in a Flash: A Further Investigation of the Photo-Taking-Impairment Effect." *Journal of Applied Research in Memory and Cognition 7*, no. 1 (2018): 154–60.

Sohei Nishino's Maps Trace More Than The City, September 3, 2018; San Francisco, CA: San Francisco Museum of Modern Art). Film, 4:49. www.sfmoma.org/watch/sohei-nishinos-maps-trace-more-city/.

Solomon R. Guggenheim Museum. *Countryside, the Future*. New York, March 2020.

Sontag, Susan. *On Photography*. New York: Farrar, Straus and Giroux, 2001. Originally published in 1977.

Spirn, Anne Whiston, and Dorothea Lange. *Daring to Look: Dorothea Lange's Photographs and Reports from the Field*. Chicago: The University of Chicago Press, 2010.

Steichen, Edward. *The Family of Man*. New York: Museum of Modern Art, 1996.

Suciu, Peter. "A Photo Used to Be Worth a Thousand Words, but Thanks to Social Media Photos Have Lost Their Value." *Forbes*, October 24, 2019. www.forbes.com/sites/petersuciu/2019/10/24/a-photo-used-to-be-worth-a-thousand-words-but-thanks-to-social-media-photos-have-lost-their-value/?sh=66263cbf607f.

Szarkowski, John. *The Photographer's Eye*. New York: Museum of Modern Art, 1966.

Talbot, William Henry Fox. "The Art of Photogenic Drawing." *Photography in Print: Writings from 1816 to the Present*, 3rd ed., edited by Vicki Goldberg, 39. Albuquerque, NM: University of New Mexico Press, 1988. Originally published in 1839.

Talbot, William Henry Fox. *The Pencil of Nature*. New York: Da Capo Press, 1969. Originally published in 1844.

Talbot, William Henry Fox. *Some Account of the Art of Photogenic Drawing: Or the Process by Which Natural Objects May Be Made to Delineate Themselves Without the Aid of the Artist's Pencil*. London, England: R. and J. E. Taylor, 1839.

Tatar, Maria. *The Heroine with 1001 Faces*. New York: Liveright Publishing Corporation, 2021.

Terreform ONE (Open Network Ecology). Accessed June 16, 2020. www.terreform.org/.
Tillman, Lynne. "Stephen Shore." In *Aperture Conversations: 1985 to the Present*, edited by Melissa Harris and Michael Famighetti. New York: Aperture, 2018. Originally published in 2004.
UBC Faculty of Medicine. *The Visual Pathway*. UBC Neuroanatomy, February 21, 2014, video, 11:22. www.youtube.com/watch?app=desktop&v=TbDFrbXiz2s.
Uhlfelder, Eric. "Bloody Skies: The Fight to Reduce Deadly Bird-Plane Collisions." *National Geographic*, November 8, 2013. www.nationalgeographic.com/adventure/article/131108-aircraft-bird-strikes-faa-radar-science.
"Urban Design Toolkit: Section Two: Participation Tools." Ministry for the Environment, March 2009. https://environment.govt.nz/publications/urban-design-toolkit-third-edition/section-2-community-participation-tools/.
Venice Biennale. Take Care of Yourself. Venice, 2007.
Venturi, Robert, Denise Scott Brown, and Steven Izenour. *Learning from Las Vegas*. Cambridge, MA: The MIT Press, 1972.
Verburg, JoAnn. "Between Exposures." Essay. In *Second View: The Rephotographic Survey Project*, 5–10. Albuquerque, NM: The University of New Mexico Press, 1984.
Vienne, Véronique. "Philip Gefter." In *Aperture Conversations: 1985 to the Present*, edited by Melissa Harris and Michael Famighetti. New York: Aperture, 2018. Originally published in 2001.
Waldheim, Charles, and Andrea Hansen. *Composite Landscapes: Photomontage and Landscape Architecture*. Ostfildern, Germany: Hatje Cantz Verlag, 2014.
Weston, Edward. "Seeing Photographically." *The Complete Photographer* 9, no. 49: 1943.
Weston, Edward. "Seeing Photographically." In *The Photography Reader*, edited by Liz Wells. London, England: Routledge, 2002. Originally published in 1943.
Whyte, William H. *The Social Life of Small Urban Spaces*. New York: Project for Public Spaces, 1980.
Wolfe, Jeremy. "Photography Changes What and How Much We Remember." In *Photography Changes Everything*, edited by Marvin Heiferman. New York: Aperture, 2012.
"XX – A Miller Prize Proposal." Agency Landscape + Planning, 2019. https://agencylp.com/projects/xx-a-miller-prize-proposal/.
Zweite, Armin. "Bernd and Hilla Becher's 'Suggestions for a Way of Seeing': Ten Key Ideas." In *Typologies*, edited by Bernd Becher and Hilla Becher, 7–35. Cambridge, MA: MIT, 2004.

Index

Adams, Robert 105, 108
annotations 154–5

Bajac, Quentin 35
Balmori, Diana 38
Barret, Terry 9
Barthes, Roland 9
Baudelaire, Charles 139
Becher, Bernd and Hilla 100–1, 103n7
Bey, Dawoud 121–3, 133
burnout 140

Calle, Sophie 144–7
captions 149–50, 153–4
cards 140
Christenberry, William 126–8
Class Pictures (Bey) 121–4
communities 120–5, 144
comparison 112–14
composition 104–10
compositional sets 60–6
Cullen, Gordon 101
Cypress Swamp (Misrach) 150

Deal, Joe 106, 108–10
Dean, Tacita 132
Debord, Guy 97, 139
depth of field 14–15
design 24–5
Diorama Maps (Nishino) 134
distance 64–5
Draplin, Aaron 21
Düsseldorf School 100

Eno, Brian 140–1
exposure 14–15

Fading Away (Robinson) 92
film 39–40, 41n4
5 second rule 31–2, 41, 43
Flaneur 139
focal points 62, 99, 101
Fox, William L. 115
framing: angles 100; and combining photographs 131; methods 52–4, 56, 57–8, 60; portraits 123; and seeing 36
Frazier, Latoya Ruby 85–6, 151

games 137–41
Goldberg, Roselee 80
Green Warehouse, Newbern, Alabama (Christenberry) 127
Green, Naima 120, 123–4

Halprin, Lawrence 101
Helphand, Kenny 153
Hockney, David 130–4
Hops (Helphand) 154
horizon line 60–1

I giardini/The Gardens (Wall) 92
I Remember Pittsburgh (Michals) 151–2
Image of the City (Lynch) 97

Jenkins, William 92–3
Jewels From the Hinterland (Green) 123–4

Klett, Mark 114–17

landscape architecture: and photography 1–4, 8; purpose of 10, 121
landscapes: changes in 137–8; designing 3; examining 1; nature of 130
Lange, Dorothea 119
Lê, An-My 45
lenses 36
limits 35–40
looking 48–50
Low, Setha 19n9
Lynch, Kevin 97

Man Going to Heaven, A (Michals) 91
Maniatis, Efthimios 13–14
Mann, Sally 54
Matthews, Laurie 113–14
measuring 13–14, 100
methods 13–15
Michals, Duane 91–3, 151–2
Misrach, Richard 149–50
missing images 84–6
motivations 43–4
movement 53–6, 138–9

narratives 89–95, 149–56
Nature of Photographs, The (Shore) 77
near and far 64–5
New Objectivity 100
New Topographics 106–8, 109
note-taking 76–9, 114
Notion of Family, The (Frazier) 151

Oblique Strategies (Eno and Schmidt) 140–1
On Photography (Sontag) 139
Opie, Catharine 29–30

Paradise Regained (Michals) 91–2
Pearlblossom Highway (Hockney) 132
Perloff, Stephen 93
Petrochemical America (Misrach) 149–50
phones 22, 36
photographs: combining 130–5; displaying 100, 126; evaluating 143–7; exhibiting 108; and narratives 149–55; *see also specific works*
photography: black and white 104; importance of 1–2; landscape 106–8, 114; meaning 8–12; mental 48–50; methods 27; practices 87; as process 4–7, 13–16, 30; purpose of 22–4, 44–5, 84, 91–2, 120, 123; re-photography 112–19, 126, 129; as representational medium 38; site 21; as skill 2–4; understanding of 80
places 120–1, 126–9, 143
point of view 67–75, 104–9, 136
portraits 120–5, 151
post-production 58
practices 87
psychogeography 139

questions 43–7

Red Building in Forest, Hale County, Alabama, The (Christenberry) 127
regularized photographing 97–102
relatability 25
Rephotographic Survey Project (RSP) 114–17
re-photography 112–19, 126, 129
resonance 25
Robinson, Henry Peach 92–3
rule of fourths 62
rule of thirds 61–2

scanning 52–4
Schmidt, Peter 140
seeing: and perspective 72–5, 101; vs. photographing 7–8, 29, 36; process of 136n3
semiotics 20
Sherald, Amy 123
Shiff, Richard 20
Shore, Stephen 54, 77
sight *see* seeing
sites 20–1
Situationist International (SI) 139
Sohei Nishino 134–5
Sontag, Susan 139
Spatializing Culture (Low) 19n9
storyboards 89–90
Street Portraits (Bey) 133
subjects 24
systems 68–9

Take Care of Yourself (Calle) 144–7
themes 44
time 29–30, 80–2, 126–9
Time and More, Space and More (Hockney) 132
transects 97

Underground Club, Greensboro Alabama (Christenberry) 127

Verburg, JoAnn 116
vernacular 108

Walking in the Zen Garden at the Ryoanji Temple (Hockney) 133
Wall, Jeff 49, 92
West and West (Deal) 108–9
Winogrand, Garry 91
writing 151–2